UGLY'S™

Electric Motors and Controls

Printed in the U.S.A.

A note from the author . . .

Ugly's Electric Motors and Controls is based on the 2008 NEC® and is designed to be used as a quick on-the-job reference in the electrical industry. We have tried to include the most commonly required information in an easy-to-read format.

We salute the National Fire Protection Association for their dedication to the protection of lives and property from fire and electrical hazards through the sponsorship of the *National Electrical Code®*.

NATIONAL ELECTRICAL CODE® and *NEC®* are registered trademarks of the National Fire Protection Association, Inc., Quincy, MA.

JONES AND BARTLETT PUBLISHERS
Sudbury, Massachusetts
BOSTON   TORONTO   LONDON   SINGAPORE

TABLE OF CONTENTS

THE LEFT-HAND RULE	1
MAGNETIC FIELD AROUND A COIL	2
ELECTROMAGNETIC INDUCTION	3
THE LEFT-HAND RULE FOR GENERATORS	4
BASIC MOTOR OPERATION	5
BASIC DC MOTOR: SPEED CONTROL	6
CALCULATING SYNCHRONOUS SPEED	7
CALCULATING RUNNING SPEED	7
AC MOTOR OPERATION AT OVER- AND UNDER-VOLTAGES	8
DC MOTOR OPERATION AT OVER- AND UNDER-VOLTAGES	9
MINIMUM DEPTH OF CLEAR WORKING SPACE IN FRONT OF ELECTRICAL EQUIPMENT	10
MINIMUM CLEARANCE OF LIVE PARTS	11
HORSEPOWER RATINGS FOR NEMA STARTERS	12
MAXIMUM HORSEPOWER	13
RUNNING OVERLOAD UNITS	14
MOTOR BRANCH-CIRCUIT PROTECTIVE DEVICES MAXIMUM RATING OR SETTING	15
FULL-LOAD CURRENT FOR DIRECT-CURRENT MOTORS IN AMPERES	16
FULL-LOAD CURRENT FOR SINGLE-PHASE ALTERNATING CURRENT MOTORS IN AMPERES	17
THREE-PHASE ALTERNATING CURRENT MOTORS FULL-LOAD CURRENT	18
FULL-LOAD CURRENT AND OTHER DATA FOR THREE-PHASE AC MOTORS	19
MOTOR AND MOTOR CIRCUIT CONDUCTOR PROTECTION	20
GENERAL MOTOR RULES	20
MOTOR BRANCH-CIRCUIT AND FEEDER EXAMPLE	21
GENERAL MOTOR APPLICATIONS	21
MAXIMUM MOTOR LOCKED-ROTOR CURRENT	22
MAXIMUM MOTOR LOCKED-ROTOR CURRENT IN AMPERES, TWO & THREE PHASE, DESIGN B, C, AND D	22
OHM'S LAW	23
ELECTRICAL FORMULAS FOR CALCULATING AMPERES, HORSEPOWER, KILOWATTS, AND KVA	25
TO FIND AMPERES	26
TO FIND HORSEPOWER	30
TO FIND WATTS	31
TO FIND KILOWATTS	32
TO FIND KILOVOLT-AMPERES	33
KIRCHHOFF'S LAWS	33

9001040393

TABLE OF CONTENTS (continued)

TO FIND INDUCTANCE	34
INDUCTANCE (L)	34
TO FIND IMPEDANCE	35
IMPEDANCE (Z)	35
TO FIND REACTANCE	36
REACTANCE (X)	36
SYNCHRONOUS SPEED	37
SLIP	37
LOCKED-ROTOR CURRENT THREE PHASE	37
LOCKED-ROTOR CURRENT SINGLE PHASE	38
MOTOR FRAMES	38
FRAME DIMENSIONS, IN INCHES	39
FRAME LETTER DESIGNATIONS	40
INSULATION CLASSES	41
COMMON ELECTRICAL DISTRIBUTION SYSTEMS	42
VOLTAGE DROP CALCULATIONS, INDUCTANCE NEGLIGIBLE	44
SINGLE-PHASE CIRCUITS	44
THREE-PHASE CIRCUITS	44
VOLTAGE DROP CALCULATION EXAMPLES	45
SINGLE-PHASE VOLTAGE DROP	47
THREE-PHASE VOLTAGE DROP	47
MAXIMUM PERMISSIBLE CAPACITOR KVAR	48
FOR USE WITH OPEN-TYPE THREE-PHASE SIXTY-CYCLE INDUCTION MOTORS	48
POWER-FACTOR CORRECTION	49
POWER FACTOR AND EFFICIENCY EXAMPLE	50
LOCKED-ROTOR CODE LETTERS	52
SINGLE-PHASE MOTORS	53
THREE-PHASE MOTORS	53
THREE-PHASE AC MOTOR WINDINGS AND CONNECTIONS	54
NEMA ENCLOSURE TYPES NONHAZARDOUS LOCATIONS	55
NEMA ENCLOSURE TYPES HAZARDOUS LOCATIONS	58
U.S. WEIGHTS AND MEASURES	59
LINEAR MEASURES	59
MILE MEASUREMENTS	59
OTHER LINEAR MEASUREMENTS	59
SQUARE MEASURES	60
CUBIC OR SOLID MEASURES	60
LIQUID MEASUREMENTS	61
DRY MEASURES	61
WEIGHT MEASUREMENTS (MASS)	62
METRIC SYSTEM	63
PREFIXES	63

TABLE OF CONTENTS (continued)

LINEAR MEASURES	63
SQUARE MEASURES	63
CUBIC MEASURES	64
MEASURES OF WEIGHT	65
MEASURES OF CAPACITY	66
METRIC DESIGNATOR AND TRADE SIZES	67
U.S. WEIGHT AND MEASURES/METRIC EQUIVALENT CHART	67
EXPLANATION OF SCIENTIFIC NOTATION	67
USEFUL CONVERSIONS/EQUIVALENTS	68
DECIMAL EQUIVALENTS	69
SINGLE-PHASE MOTORS	71
DIRECT-CURRENT MOTORS	73
MOTOR SELECTION CHECKLIST	74
MOTOR SELECTION CRITERIA	75
TYPICAL LOAD SERVICE FACTORS	76
OVERCURRENT PROTECTION FOR TWO OR MORE MOTORS	77
OVERCURRENT PROTECTION FOR MOTORS AND OTHER LOADS	78
DETERMINING OVERLOAD SIZE	79
DETERMINING CONTROLLER SIZE	80
MOTORS 2 HORSEPOWER OR LESS AND 300 VOLTS OR LESS	81
DETERMINING CONDUCTOR SIZES FOR SINGLE-PHASE MOTORS	82
DETERMINING CONDUCTOR SIZES FOR ADJUSTABLE SPEED DRIVES	83
MOTOR AND MOTOR-CIRCUIT CONDUCTOR PROTECTION	83
GENERAL MOTOR RULES	84
MOTOR BRANCH CIRCUIT AND FEEDER EXAMPLE	84
GENERAL MOTOR APPLICATIONS	84
APPROXIMATE TORQUE FIGURES, COMPOUND DC MOTORS	86
HEATER CORRECTIONS FOR AMBIENT TEMPERATURES	86
APPROXIMATE TORQUE FIGURES, AC MOTORS	87
APPROXIMATE TORQUE FIGURES, WOUND ROTOR MOTORS	88
SIZING LOAD CONDUCTORS FROM GENERATORS	89
ELECTRICAL SAFETY DEFINITIONS	90
ELECTRICAL SAFETY CHECKLIST	93
ELECTRICAL SAFETY LOCKOUT–TAGOUT PROCEDURES	94
APPLICATION OF LOCKOUT–TAGOUT DEVICES	94
REMOVAL OF LOCKOUT–TAGOUT DEVICES	95
ELECTRICAL SAFETY SHOCK PROTECTION BOUNDARIES	96
ELECTRICAL SAFETY HOW TO READ A WARNING LABEL	97
PULLEY CALCULATIONS	98
DETERMINING BELT LENGTH	99

TABLE OF CONTENTS (continued)

```
HORSEPOWER CAPACITIES .............................. 99
GEAR SIZING ........................................100
DETERMINING SHAFT DIAMETER ........................100
GEAR REDUCERS .....................................101
        OUTPUT TORQUE .............................101
        OUTPUT SPEED ..............................101
        OUTPUT HORSEPOWER .........................101
MOTOR TORQUE ......................................102
        TORQUE ....................................102
        STARTING TORQUE ...........................102
CALCULATING COST OF OPERATING AN ELECTRICAL
    APPLIANCE .....................................102
ELECTRICAL SYMBOLS ................................103
WIRING DIAGRAMS ...................................105
COMPLETE STOP-START SYSTEM WITH CONTROL
    TRANSFORMER ...................................108
HAND OFF AUTOMATIC CONTROL ........................109
JOGGING WITH CONTROL RELAY ........................110
WIRING DIAGRAMS ...................................111
PLUGGING CIRCUIT ..................................113
TERMINAL DESIGNATIONS .............................114
        GENERATORS AND SYNCHRONOUS MOTORS .........114
COUNTER-EMF STARTING ..............................115
        DC MOTORS .................................115
TWO-SPEED STARTING ................................116
REDUCED-VOLTAGE STARTING ..........................117
        SYNCHRONOUS MOTOR .........................117
```

THE LEFT-HAND RULE

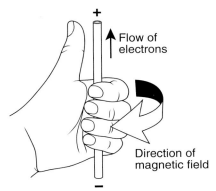

Any time an electrical current flows through a conductor, it creates a magnetic field around it. The left-hand rule identifies the directions of the current and the magnetic field. If you hold the conductor in your left hand, with your thumb pointing from the negative to the positive pole, the magnetic field around the conductor will always be in the direction your fingers point.

MAGNETIC FIELD AROUND A COIL

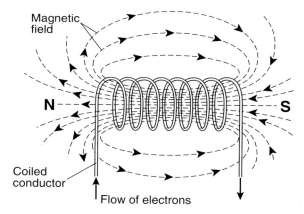

If a conductor is wound into a coil, the magnetic fields of each turn add together, producing a very strong field.

ELECTROMAGNETIC INDUCTION

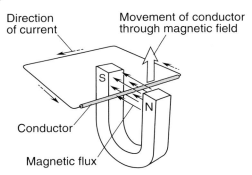

Electromagnetic induction occurs when energy is transferred in the form of magnetic fields, with no physical connection between circuits. It is the result of relative movement between a conductor and a magnetic field. Whether the conductor moves or the field moves, there is no difference. The only critical issue is that the conductor moves through the magnetic flux.

THE LEFT-HAND RULE FOR GENERATORS

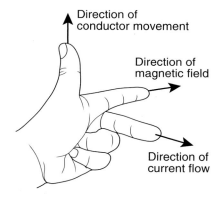

A second type of left-hand rule is used to identify the motions of conductors, currents, and fields in a generator. To do this, position your index finger, thumb, and middle finger to be perpendicular (at 90° angles) to one another. (See drawing.) Turn your hand so that your index finger points in the direction of the magnetic field and your thumb in the direction of conductor movement. Your middle finger will be pointing in the direction of current flow.

BASIC MOTOR OPERATION

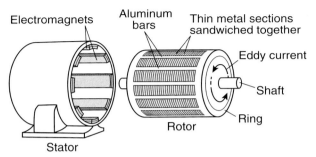

A motor's *rotor* is built from thin metal sections so that eddy currents are reduced. These sections have embedded bars that are welded together with a ring. Current travels through the bars of the rotor.

The poles of the *stator* create a powerful magnetic field that rises and falls with each alternation of current. The rotor is either attracted or repelled by the stator poles. By cleverly arranging the positions of the poles, rotor, and current, the motor is kept moving in one direction.

BASIC DC MOTOR: SPEED CONTROL

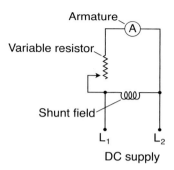

The resistor reduces the current through the armature, and with it, the speed of the motor.

CALCULATING SYNCHRONOUS SPEED

Synchronous Speed = $\dfrac{120 \times \text{Frequency}}{\text{Number of poles}}$

Example: $\dfrac{120 \times 60 \text{ Hz}}{6 \text{ poles}}$ = 7200 ÷ 6 = 1200 rpm

CALCULATING RUNNING SPEED

Running Speed = Synchronous Speed − Slip

Example: If slip = 5%,
　　　　running speed = synchronous speed − 5%, or
　　　　　　　　　　　　95% of synchronous speed.

　　　　1200 × 0.95 = 1140 rpm

AC MOTOR OPERATION AT OVER- AND UNDER-VOLTAGES*

	At 10% Under-Voltage
Efficiency	−2%
Temperature	+18%
Current (full load)	+12%
Speed at full load	+1%
Torque	−20%
Power factor	−4%

	At 10% Over-Voltage
Efficiency	+1%
Temperature	−9%
Current (full load)	−7%
Speed at full load	−2%
Torque	+20%
Power factor	+4%

All figures approximate.

DC MOTOR OPERATION AT OVER- AND UNDER-VOLTAGES

Characteristic	At 10% Under-Voltage		At 10% Over-Voltage	
	Shunt	Compound	Shunt	Compound
Speed	−5%	−6%	+5%	+6%
Current	+12%	+12%	−8%	−8%
Shunt Torque	−15%	−15%	+15%	+15%
Field Temp	Increased	Decreased	Increased	Increased
Commutator Temp	Increased	Increased	Decreased	Decreased
Armature Temp	Increased	Increased	Decreased	Decreased

MINIMUM DEPTH OF CLEAR WORKING SPACE IN FRONT OF ELECTRICAL EQUIPMENT

NOMINAL VOLTAGE TO GROUND	CONDITIONS		
	1	2	3
	Minimum clear distance (feet)		
0 - 150	3	3	3
151 - 600	3	3½	4
601 - 2500	3	4	5
2501 - 9000	4	5	6
9001 - 25,000	5	6	9
25,001 - 75 kV	6	8	10
Above 75 kV	8	10	12

NOTES:
  Condition 1 *= Live parts are exposed on one side of the working space only. Or: Live parts are exposed on both sides of the working space, but are all sufficiently guarded by insulating materials.*
  Condition 2 *= Live parts are exposed on one side of the working space and grounded parts are exposed on the other side of the working space. Walls are considered to be grounded if made of concrete, brick, or tile.*
  Condition 3 *= Live parts are exposed on both sides of the working space.*

Adapted from *NEC*® Tables 110.26(A)(1) and 110.34(A).

MINIMUM CLEARANCE OF LIVE PARTS

NOMINAL VOLTAGE RATING KV	IMPULSE WITHSTAND B.I.L. KV		MINIMUM CLEARANCE OF LIVE PARTS, INCHES			
			PHASE-TO-PHASE		PHASE-TO-GROUND	
	OUTDOORS	INDOORS	OUTDOORS	INDOORS	OUTDOORS	INDOORS
2.4 - 4.16	95	60	7	4.5	6	3.0
7.2	95	75	7	5.5	6	4.0
13.8	110	95	12	7.5	7	5.0
14.4	110	110	12	9.0	7	6.5
23	150	125	15	10.5	10	7.5
34.5	150	150	15	12.5	10	9.5
	200	200	18	18.0	13	13.0
46	200	–	18	–	13	–
	250	–	21	–	17	–
69	250	–	21	–	17	–
	350	–	31	–	25	–
115	550	–	53	–	42	–
138	550	–	53	–	42	–
	650	–	63	–	50	–
161	650	–	63	–	50	–
	750	–	72	–	58	–
230	750	–	72	–	58	–
	900	–	89	–	71	–
	1050	–	105	–	83	–

- The clearances listed above are for rigid parts and bare conductors, under what the NEC calls "favorable service conditions." Where there will be conductor movement or other complications will be present, the above clearances should be exceeded. Extra space should always be provided if possible.
- The impulse-withstand voltage is determined by a system's surge protective equipment.

Adapted from NEC® Table 490.24.

HORSEPOWER RATINGS FOR NEMA STARTERS

NEMA Size	Volts (VAC)	Single Phase	Three Phase	Current Rating (Amperes)
00	115	1/3	—	9
	200	—	1½	
	230	1	1½	
	460/575	—	2	
0	115	1	—	18
	200	—	3	
	230	2	3	
	460/575	—	5	
1	115	2	—	27
	200	—	7½	
	230	3	7½	
	460/575	—	10	
1P	115	3	—	36
	230	5	—	
2	115	3	—	45
	200	—	10	
	230	7½	15	
	460/575	—	25	
3	115	7½	—	90
	200	—	25	
	230	15	30	
	460/575	—	50	
4	200	—	40	135
	230	—	50	
	460/575	—	100	
5	200	—	75	270
	230	—	100	
	460/575	—	200	
6	200	—	150	540
	230	—	200	
	460/575	—	400	
7	230	—	300	90
	460/575	—	600	
8	230	—	450	1215
	460/575	—	400	
9	460/575	—	900	2250
	230	—	800	

Reprinted from Miller, Charles R. *NFPA's Pocket Electrical References, First Edition.* Jones and Bartlett Publishers, 2007.

MAXIMUM HORSEPOWER

Three-Phase Motors

NEMA Size	Full-Voltage Starting			Auto-Transformer Starting			Part-Winding Starting			Wye-Delta Starting		
	200 VAC	230 VAC	460/575 VAC	200 VAC	230 VAC	460/575 VAC	200 VAC	230 VAC	460/575 VAC	200 VAC	230 VAC	460/575 VAC
00	1½	1½	2	—	—	—	—	—	—	—	—	—
0	3	3	5	—	—	—	—	—	—	—	—	—
1	7½	7½	10	7½	7½	10	10	10	15	10	10	15
2	10	15	25	10	15	25	20	25	40	20	25	40
3	25	30	50	25	30	50	40	50	75	40	50	75
4	40	50	100	40	50	100	75	75	150	60	75	150
5	75	100	200	75	100	200	150	150	350	150	150	300
6	150	200	400	150	200	400	—	300	600	300	350	700
7	—	300	600	—	300	600	—	450	900	500	500	1000
8	—	450	900	—	450	900	—	700	1400	750	800	1500
9	—	800	1600	—	800	1600	—	1300	2600	1500	1500	3000

Reprinted from Miller, Charles R. *NFPA's Pocket Electrical References, First Edition.* Jones and Bartlett Publishers, 2007.

RUNNING OVERLOAD UNITS

TYPE OF MOTOR	SUPPLY	NUMBER AND LOCATION OF OVERLOAD UNITS
1-phase AC or DC	• 2-wire • 1-phase AC or DC • ungrounded	1 overload in either conductor
1-phase AC or DC	• 2-wire • 1-phase AC or DC • one conductor ungrounded	1 overload in ungrounded conductor
1-phase AC or DC	• 3-wire • 1-phase AC or DC • grounded neutral conductor	1 overload in either ungrounded conductor
1-phase AC	• any 3-phase	1 overload in ungrounded conductor
2-phase AC	• 3-wire • 2-phase AC • ungrounded	2 overloads, one in each phase
2-phase AC	• 3-wire • 2-phase AC • one conductor grounded	2 overloads in ungrounded conductors
2-phase AC	• 4-wire • 2-phase AC • grounded or ungrounded	2 overloads, one per phase in ungrounded conductors
2-phase AC	• 5-wire • 2-phase AC • grounded neutral or ungrounded	2 overloads, one per phase in any ungrounded phase wire
3-phase AC	• any 3-phase	3 overloads, one in each phase*

* Exception: Where protected by other approved means.

Adapted from *NEC*® Table 430.37.

MOTOR BRANCH-CIRCUIT PROTECTIVE DEVICES MAXIMUM RATING OR SETTING

Type of Motor	Nontime Delay Fuse	Dual-Element (Time-Delay) Fuse	Instantaneous Trip Breaker	Inverse Time Breaker
AC polyphase motors other than wound rotor	300% FLC	175% FLC	800% FLC	250% FLC
DC (constant voltage)	150% FLC	150% FLC	250% FLC	150% FLC
Design B energy-efficient	300% FLC	175% FLC	1100% FLC	250% FLC
Single-phase motors	300% FLC	175% FLC	800% FLC	250% FLC
Squirrel cage—other than Design B energy-efficient	300% FLC	175% FLC	800% FLC	250% FLC
Synchronous	300% FLC	175% FLC	800% FLC	250% FLC
Wound rotor	150% FLC	150% FLC	800% FLC	150% FLC

- There are exceptions to this chart. See *NEC* Sections 430.52–430.54.
- Nontime Delay Fuse = Class CC fuses.
- For nonadjustable inverse time circuit breakers, see *NEC* Section 430.52.
- For low-torque, low-speed, synchronous motors that drive reciprocating compressors, pumps, or similar loads, and which start unloaded, fuse or circuit-breaker settings need not exceed 200% of full-load current.

Adapted from *NEC*® Table 430.52.

FULL-LOAD CURRENT FOR DIRECT-CURRENT MOTORS IN AMPERES

Average DC Quantity HP	Armature Voltage Rating for Motors Running at Base Speed				
	90V	120V	180V	240V	500V
1/4	4.0	3.1	2.0	1.6	–
1/3	5.2	4.1	2.6	2.0	–
1/2	6.8	5.4	3.4	2.7	–
3/4	9.6	7.6	4.8	3.8	–
1	12.2	9.5	6.1	4.7	–
1½	–	13.2	8.3	6.6	–
2	–	17	10.8	8.5	–
3	–	25	16	12.2	–
5	–	40	27	20	–
7½	–	58	–	29	13.6
10	–	76	–	38	18
15	–	–	–	55	27
20	–	–	–	72	34
25	–	–	–	89	43
30	–	–	–	106	51
40	–	–	–	140	67
50	–	–	–	173	83
60	–	–	–	206	99
75	–	–	–	255	123
100	–	–	–	341	164
125	–	–	–	425	205
150	–	–	–	506	246
200	–	–	–	675	330

Adapted from *NEC*® Table 430.247.

FULL-LOAD CURRENT FOR SINGLE-PHASE ALTERNATING CURRENT MOTORS IN AMPERES

HP	115V	200V	208V	230V
1/4	5.8	3.3	3.2	2.9
1/3	7.2	4.1	4.0	3.6
1/2	9.8	5.6	5.4	4.9
3/4	13.8	7.9	7.6	6.9
1	16	9.2	8.8	8.0
1½	20	11.5	11	10
2	24	13.8	13.2	12
3	34	19.6	18.7	17
5	56	32.2	30.8	28
7½	80	46	44	40
10	100	57.5	55	50

The voltages listed are rated motor voltages. The listed currents are for system voltage ranges of 110 to 120 and 220 to 240.

Adapted from *NEC*® Table 430.248.

THREE-PHASE ALTERNATING CURRENT MOTORS FULL-LOAD CURRENT

HP	Induction Type Squirrel-Cage and Wound-Rotor Amperes						Synchronous Type Unity Power Factor* Amperes			
	115V	200V	208V	230V	460V	575V	2300V	230V	460V	575V	2300V
1/2	4.4	2.5	2.4	2.2	1.1	0.9	–	–	–	–	–
3/4	6.4	3.7	3.5	3.2	1.6	1.3	–	–	–	–	–
1	8.4	4.8	4.6	4.2	2.1	1.7	–	–	–	–	–
1 1/2	12.0	6.9	6.6	6.0	3.0	2.4	–	–	–	–	–
2	13.6	7.8	7.5	6.8	3.4	2.7	–	–	–	–	–
3	–	11.0	10.6	9.6	4.8	3.9	–	–	–	–	–
5	–	17.5	16.7	15.2	7.6	6.1	–	–	–	–	–
7 1/2	–	25.3	24.2	22	11	9	–	–	–	–	–
10	–	32.2	30.8	28	14	11	–	–	–	–	–
15	–	48.3	46.2	42	21	17	–	–	–	–	–
20	–	62.1	59.4	54	27	22	–	–	–	–	–
25	–	78.2	74.8	68	34	27	–	53	26	21	–
30	–	92	88	80	40	32	–	63	32	26	–
40	–	120	114	104	52	41	–	83	41	33	–
50	–	150	143	130	65	52	–	104	52	42	–
60	–	177	169	154	77	62	16	123	61	49	12
75	–	221	211	192	96	77	20	155	78	62	15
100	–	285	273	248	124	99	26	202	101	81	20
125	–	359	343	312	156	125	31	253	126	101	25
150	–	414	396	360	180	144	37	302	151	121	30
200	–	552	528	480	240	192	49	400	201	161	40
250	–	–	–	–	302	242	60	–	–	–	–
300	–	–	–	–	361	289	72	–	–	–	–
350	–	–	–	–	414	336	83	–	–	–	–
400	–	–	–	–	477	382	95	–	–	–	–
450	–	–	–	–	515	412	103	–	–	–	–

The voltages listed are rated motor voltages. The currents listed are for system voltage ranges of 110 to 120, 220 to 240, 440 to 480, and 550–600 volts.

* Multiply by 1.1 and 1.25 for 90- and 80-percent power factor, respectively.

Adapted from *NEC*® Table 430.250.

FULL-LOAD CURRENT AND OTHER DATA FOR THREE-PHASE AC MOTORS

MOTOR HORSEPOWER		MOTOR AMPERE	SIZE BREAKER	SIZE STARTER	HEATER AMPERE **	SIZE WIRE	SIZE CONDUIT
½	230V	2.2	15	00	2.530	12	¾"
	460	1.1	15	00	1.265	12	¾"
¾	230	3.2	15	00	3.680	12	¾
	460	1.6	15	00	1.840	12	¾
1	230	4.2	15	00	4.830	12	¾
	460	2.1	15	00	2.415	12	¾
1½	230	6.0	15	00	6.900	12	¾
	460	3.0	15	00	3.450	12	¾
2	230	6.8	15	0	7.820	12	¾
	460	3.4	15	00	3.910	12	¾
3	230	9.6	20	0	11.040	12	¾
	460	4.8	15	0	5.520	12	¾
5	230	15.2	30	1	17.480	12	¾
	460	7.6	15	0	8.740	12	¾
7½	230	22	45	1	25.300	10	¾
	460	11	20	1	12.650	12	¾
10	230	28	60	2	32.200	10	¾
	460	14	30	1	16.100	12	¾
15	230	42	70	2	48.300	6	1
	460	21	40	2	24.150	10	¾
20	230	54	100	3	62.100	4	1
	460	27	50	2	31.050	10	¾
25	230	68	100	3	78.200	4	1½
	460	34	50	2	39.100	8	1
30	230	80	125	3	92.000	3	1½
	460	40	70	3	46.000	8	1
40	230	104	175	4	119.600	1	1½
	460	52	100	3	59.800	6	1
50	230	130	200	4	149.500	00	2
	460	65	150	3	74.750	4	1½

* Overcurrent device may have to be increased due to starting current and load conditions. See *NEC* 430–52, Table 430–52. Wire size based on 75°C terminations and 75°C insulation.

** Overload heater must be based on motor nameplate & sized per *NEC* 430–32.

*** Conduit size based on Rigid Metal Conduit with some spare capacity. For minimum size & other conduit types, see *NEC* Appendix C.

FULL-LOAD CURRENT AND OTHER DATA FOR THREE-PHASE AC MOTORS

MOTOR HORSEPOWER		MOTOR AMPERE	SIZE BREAKER	SIZE STARTER	HEATER AMPERE **	SIZE WIRE	SIZE CONDUIT
60	230V	154	250	5	177.10	000	2"
	460	77	200	4	88.55	3	1½
75	230	192	300	5	220.80	250 kcmil	2½
	460	96	200	4	110.40	1	1½
100	230	248	400	5	285.20	350 kcmil	3
	460	124	200	4	142.60	2/0	2
125	230	312	500	6	358.80	600 kcmil	3½
	460	156	250	5	179.40	000	2
150	230	360	600	6	414.00	700 kcmil	4
	460	180	300	5	207.00	0000	2½

Motor and Motor Circuit Conductor Protection

Motors can have large starting currents three to five times or more than that of the actual motor current. In order for motors to start, the motor and motor circuit conductors are allowed to be protected by circuit breakers and fuses at values that are higher than the actual motor and conductor ampere ratings. These larger overcurrent devices do not provide overload protection and will only open upon short circuits or ground faults. Overload protection must be used to protect the motor based on the actual nameplate amperes of the motor. This protection is usually in the form of heating elements in manual or magnetic motor starters. Small motors such as waste disposal motors have a red overload reset button built into the motor.

General Motor Rules

- Use Full-Load Current from tables instead of nameplate.
- Branch-Circuit Conductors - Use 125% of Full-Load Current to find conductor size.
- Branch-Circuit OCP Size - Use percentages given in tables for Full-Load Current.
- Feeder Conductor Size - 125% of largest motor and sum of the rest.
- Feeder OCP - Use largest OCP plus rest of Full-Load Currents

MOTOR BRANCH-CIRCUIT AND FEEDER EXAMPLE

General Motor Applications

Branch-Circuit Conductors: Use Full-Load Three-Phase Currents; From *NEC* Table 430.250, 50 HP 480 volt Three-Phase motor design B, 75 degree terminations = 65 Amperes
125% of Full-Load Current [*NEC* 430.22(A)] 125% of 65 A = **81.25 Amperes** Conductor Selection Ampacity

Branch-Circuit Overcurrent Device: *NEC* 430.52 (C1)
(Branch-Circuit Short Circuit and Ground-Fault Protection)
Use percentages given in *NEC* 430.52 for **Type** of circuit breaker or fuse used.
50 HP 480 V 3 Ph Motor = 65 Amperes.
Nontime Fuse = 300%.
300% of 65A = 195 A. *NEC* 430.52(C1)(EX1) Next size allowed *NEC* 240.6A = **200 Ampere Fuse**.

Feeder Connectors: For 50 HP and 30 HP 480 Volt Three-Phase design B motors on same feeder
Use 125% of largest full-load current and 100% of rest. (*NEC* 430.24)
50 HP 480 V 3 Ph Motor = 65A; 30 HP 480 V 3 Ph Motor = 40A
(125% of 65A) + 40A = **121.25 A** Conductor Selection Ampacity

Feeder Overcurrent Device: *NEC* 430.62(A)
(Feeder short circuit and ground-fault protection)
Use largest overcurrent protection device <u>plus</u> full-load currents of the rest of the motors.
50 HP = 200 A fuse (65 FLC)
30 HP = 125 A fuse (40 FLC)
200 A fuse + 40 A (FLC) = 240 A. Do not exceed this value on feeder. Go down to a **225 A fuse**.

MAXIMUM MOTOR LOCKED-ROTOR CURRENT*

HP	115V	208V	230V	HP	115V	208V	230V
1/2	58.8	32.5	29.4	3	204	113	102
3/4	82.8	45.8	41.4	5	336	186	168
1	96	53	48	7½	480	265	240
1½	120	66	60	10	600	332	300
2	144	80	72				

Adapted from *NEC*® Table 430.251(A).
* Conversion Table for Selection of Disconnecting Means and Controllers as Determined from Horsepower and Voltage Rating. For use only with 430.110, 440.12, 440.41 and 455.8(C).

MAXIMUM MOTOR LOCKED-ROTOR CURRENT IN AMPERES, TWO & THREE PHASE, DESIGN B, C, AND D **

HP	115V	200V	208V	230V	460V	575V
1/2	40	23	22.1	20	10	8
3/4	50	28.8	27.6	25	12.5	10
1	60	34.5	33	30	15	12
1½	80	46	44	40	20	16
2	100	57.5	55	50	25	20
3	–	73.6	71	64	32	25.6
5	–	105.8	102	92	46	36.8
7½	–	146	140	127	63.5	50.8
10	–	186.3	179	162	81	64.8
15	–	267	257	232	116	93
20	–	334	321	290	145	116
25	–	420	404	365	183	146
30	–	500	481	435	218	174
40	–	667	641	580	290	232
50	–	834	802	725	363	290
60	–	1001	962	870	435	348
75	–	1248	1200	1085	543	434
100	–	1668	1603	1450	725	580
125	–	2087	2007	1815	908	726
150	–	2496	2400	2170	1085	868
200	–	3335	3207	2900	1450	1160

Adapted from *NEC*® Table 430.251(B).
** Conversion Table for Selection of Disconnecting Means and Controllers as Determined from Horsepower and Voltage Rating and Design Letter. For use only with 430.110, 440.12, 440.41 and 455.8(C).

OHM'S LAW

The rate of the flow of the current is equal to electromotive force divided by resistance.

I = Intensity of Current = Amperes
E = Electromotive Force = Volts
R = Resistance = Ohms
P = Power = Watts

The three basic Ohm's Law formulas are:

$$I = \frac{E}{R} \qquad R = \frac{E}{I} \qquad E = I \times R$$

Below is a chart containing the formulas related to Ohm's Law. To use the chart, from the center circle, select the value you need to find, I (Amps), R (Ohms), E (Volts) or P (Watts). Then select the formula containing the values you know from the corresponding chart quadrant.

Example:
An electric appliance is rated at 1200 Watts, and is connected to 120 Volts. How much current will it draw?

Amperes = $\frac{\text{Watts}}{\text{Volts}}$ $\qquad I = \frac{P}{E} \qquad I = \frac{1200}{120} = 10$ A

What is the Resistance of the same appliance?

Ohms = $\frac{\text{Volts}}{\text{Amperes}}$ $\qquad R = \frac{E}{I} \qquad R = \frac{120}{10} = 12 \; \Omega$

OHM'S LAW

In the preceding example, we know the following values:
I = amps = 10   R = ohms = 12Ω
E = volts = 120   P = watts = 1200

We can now see how the twelve formulas in the Ohm's Law chart can be applied.

AMPS = $\sqrt{\dfrac{\text{WATTS}}{\text{OHMS}}}$ $I = \sqrt{\dfrac{P}{R}} = \sqrt{\dfrac{1200}{12}} = \sqrt{100} = 10A$

AMPS = $\dfrac{\text{WATTS}}{\text{VOLTS}}$ $I = \dfrac{P}{E} = \dfrac{1200}{120} = 10A$

AMPS = $\dfrac{\text{VOLTS}}{\text{OHMS}}$ $I = \dfrac{E}{R} = \dfrac{120}{12} = 10A$

WATTS = $\dfrac{\text{VOLTS}^2}{\text{OHMS}}$ $P = \dfrac{E^2}{R} = \dfrac{120^2}{12} = \dfrac{14{,}400}{12} = 1200W$

WATTS = VOLTS × AMPS $P = E \times I = 120 \times 10 = 1200W$

WATTS = AMPS² × OHMS $P = I^2 \times R = 100 \times 12 = 1200W$

VOLTS = $\sqrt{\text{WATTS} \times \text{OHMS}}$ $E = \sqrt{P \times R} = \sqrt{1200 \times 12} = \sqrt{14{,}400} = 120V$

VOLTS = AMPS × OHMS $E = I \times R = 10 \times 12 = 120V$

VOLTS = $\dfrac{\text{WATTS}}{\text{AMPS}}$ $E = \dfrac{P}{I} = \dfrac{1200}{10} = 120V$

OHMS = $\dfrac{\text{VOLTS}^2}{\text{WATTS}}$ $R = \dfrac{E^2}{P} = \dfrac{120^2}{1{,}200} = \dfrac{14{,}400}{1{,}200} = 12\Omega$

OHMS = $\dfrac{\text{WATTS}}{\text{AMPS}^2}$ $R = \dfrac{P}{I^2} = \dfrac{1200}{100} = 12\Omega$

OHMS = $\dfrac{\text{VOLTS}}{\text{AMPS}}$ $R = \dfrac{E}{I} = \dfrac{120}{10} = 12\Omega$

ELECTRICAL FORMULAS FOR CALCULATING AMPERES, HORSEPOWER, KILOWATTS, AND KVA

TO FIND	DIRECT CURRENT	ALTERNATING CURRENT		
		SINGLE PHASE	TWO PHASE-FOUR WIRE	THREE PHASE
AMPERES WHEN "HP" IS KNOWN	$\dfrac{HP \times 746}{E \times \%EFF}$	$\dfrac{HP \times 746}{E \times \%EFF \times PF}$	$\dfrac{HP \times 746}{E \times \%EFF \times PF \times 2}$	$\dfrac{HP \times 746}{E \times \%EFF \times PF \times 1.73}$
AMPERES WHEN "KW" IS KNOWN	$\dfrac{KW \times 1000}{E}$	$\dfrac{KW \times 1000}{E \times PF}$	$\dfrac{KW \times 1000}{E \times PF \times 2}$	$\dfrac{KW \times 1000}{E \times PF \times 1.73}$
AMPERES WHEN "KVA" IS KNOWN		$\dfrac{KVA \times 1000}{E}$	$\dfrac{KVA \times 1000}{E \times 2}$	$\dfrac{KVA \times 1000}{E \times 1.73}$
KILOWATTS (True Power)	$\dfrac{E \times I}{1000}$	$\dfrac{E \times I \times PF}{1000}$	$\dfrac{E \times I \times PF \times 2}{1000}$	$\dfrac{E \times I \times PF \times 1.73}{1000}$
KILOVOLT-AMPERES "KVA" (Apparent Power)		$\dfrac{E \times I}{1000}$	$\dfrac{E \times I \times 2}{1000}$	$\dfrac{E \times I \times 1.73}{1000}$
HORSEPOWER	$\dfrac{E \times I \times \%EFF}{746}$	$\dfrac{E \times I \times \%EFF \times PF}{746}$	$\dfrac{E \times I \times \%EFF \times PF \times 2}{746}$	$\dfrac{E \times I \times \%EFF \times PF \times 1.73}{746}$

PERCENT EFFICIENCY = %EFF = $\dfrac{\text{OUTPUT (WATTS)}}{\text{INPUT (WATTS)}}$ POWER FACTOR = PF = $\dfrac{\text{POWER USED (WATTS)}}{\text{APPARENT POWER}} = \dfrac{KW}{KVA}$

E = VOLTS
I = AMPERES
W = WATTS

NOTE: DIRECT CURRENT FORMULAS DO NOT USE (PF, 2, OR 1.73)
SINGLE PHASE FORMULAS DO NOT USE (2, OR 1.73)
TWO PHASE - FOUR WIRE FORMULAS DO NOT USE (1.73)
THREE PHASE FORMULAS DO NOT USE (2)

 **TO FIND AMPERES**

DIRECT CURRENT:
A. When *HORSEPOWER* is known:

$$\text{AMPERES} = \frac{\text{HORSEPOWER} \times 746}{\text{VOLTS} \times \text{EFFICIENCY}} \quad \text{or} \quad I = \frac{\text{HP} \times 746}{\text{E} \times \text{\%EFF}}$$

What current will a travel-trailer toilet draw when equipped with a 12 volt, 1/8 HP motor, having a 96% efficiency rating?

$$I = \frac{\text{HP} \times 746}{\text{E} \times \text{\%EFF}} = \frac{746 \times 1/8}{12 \times 0.96} = \frac{93.25}{11.52} = 8.09 \text{ AMPS}$$

B. When *KILOWATTS* are known:

$$\text{AMPERES} = \frac{\text{KILOWATTS} \times 1000}{\text{VOLTS}} \quad \text{or} \quad I = \frac{\text{KW} \times 1000}{\text{E}}$$

A 75 KW, 240 Volt, direct-current generator is used to power a variable-speed conveyor belt at a rock crushing plant. Determine the current.

$$I = \frac{\text{KW} \times 1000}{\text{E}} = \frac{75 \times 1000}{240} = 312.5 \text{ AMPS}$$

SINGLE PHASE:
A. When *WATTS, VOLTS, AND POWER FACTOR* are known:

$$\text{AMPERES} = \frac{\text{WATTS}}{\text{VOLTS} \times \text{POWER FACTOR}} \quad \text{or} \quad \frac{P}{E \times PF}$$

Determine the current when a circuit has a 1500 watt load, a power factor of 86%, and operates from a single-phase 230 volt source.

$$I = \frac{1500}{230 \times 0.86} = \frac{1500}{197.8} = 7.58 \text{ AMPS}$$

TO FIND AMPERES

SINGLE PHASE:
B. When *HORSEPOWER* is known:

$$\text{AMPERES} = \frac{\text{HORSEPOWER} \times 746}{\text{VOLTS} \times \text{EFFICIENCY} \times \text{POWER FACTOR}}$$

Determine the amp load of a single-phase, 1/2 HP, 115 volt motor. The motor has an efficiency rating of 92%, and a power factor of 80%.

$$I = \frac{HP \times 746}{E \times \%EFF \times PF} = \frac{1/2 \times 746}{115 \times 0.92 \times 0.80} = \frac{373}{84.64}$$

I = 4.4 AMPS

C. When *KILOWATTS* are known:

$$\text{AMPERES} = \frac{\text{KILOWATTS} \times 1000}{\text{VOLTS} \times \text{POWER FACTOR}} \quad \text{or} \quad I = \frac{KW \times 1000}{E \times PF}$$

A 230 volt single-phase circuit has a 12KW power load, and operates at 84% power factor. Determine the current.

$$I = \frac{KW \times 1000}{E \times PF} = \frac{12 \times 1000}{230 \times 0.84} = \frac{12,000}{193.2} = 62 \text{ AMPS}$$

D. When *KILOVOLT-AMPERE* is known:

$$\text{AMPERES} = \frac{\text{KILOVOLT-AMPERE} \times 1000}{\text{VOLTS}} \quad \text{or} \quad I = \frac{KVA \times 1000}{E}$$

A 115 volt, 2 KVA, single-phase generator operating at full load will deliver 17.4 AMPERES. (Prove.)

$$I = \frac{2 \times 1000}{115} = \frac{2000}{115} = 17.4 \text{ AMPS}$$

REMEMBER:
  By definition, amperes is the rate of the flow of the current.

TO FIND AMPERES

THREE PHASE:
A. When *WATTS, VOLTS, AND POWER FACTOR are known*:

$$\text{AMPERES} = \frac{\text{WATTS}}{\text{VOLTS} \times \text{POWER FACTOR} \times 1.73}$$

or $I = \dfrac{P}{E \times PF \times 1.73}$

Determine the current when a circuit has a 1500 watt load, a power factor of 86%, and operates from a three-phase, 230 volt source.

$$I = \frac{P}{E \times PF \times 1.73} = \frac{1500}{230 \times 0.86 \times 1.73} = \frac{1500}{342.2}$$

I = 4.4 AMPS

B. When *HORSEPOWER* is known:

$$\text{AMPERES} = \frac{\text{HORSEPOWER} \times 746}{\text{VOLTS} \times \text{EFFICIENCY} \times \text{POWER FACTOR} \times 1.73}$$

or $I = \dfrac{HP \times 746}{E \times \%EFF \times PF \times 1.73}$

Determine the amp load of a three-phase, 1/2 HP, 230 volt motor. The motor has an efficiency rating of 92%, and a power factor of 80%.

$$I = \frac{HP \times 746}{E \times \%EFF \times PF \times 1.73} = \frac{1/2 \times 746}{230 \times .92 \times .80 \times 1.73}$$

$$= \frac{373}{293} = 1.27 \text{ AMPS}$$

TO FIND AMPERES

THREE PHASE:
C. When *KILOWATTS* are known:

$$\text{AMPERES} = \frac{\text{KILOWATTS} \times 1000}{\text{VOLTS} \times \text{POWER FACTOR} \times 1.73}$$

or $\quad I = \dfrac{\text{KW} \times 1000}{\text{E} \times \text{PF} \times 1.73}$

A 230 volt, three-phase circuit, has a 12KW power load, and operates at 84% power factor. Determine the current.

$$I = \frac{\text{KW} \times 1000}{\text{E} \times \text{PF} \times 1.73} = \frac{12{,}000}{230 \times 0.84 \times 1.73} = \frac{12{,}000}{334.24}$$

I = 36 AMPS

D. When *KILOVOLT-AMPERE* is known:

$$\text{AMPERES} = \frac{\text{KILOVOLT-AMPERE} \times 1000}{\text{E} \times 1.73} = \frac{\text{KVA} \times 1000}{\text{E} \times 1.73}$$

A 230 volt, 4 KVA, three-phase generator operating at full load will deliver 10 AMPERES. (Prove.)

$$I = \frac{\text{KVA} \times 1000}{\text{E} \times 1.73} = \frac{4 \times 1000}{230 \times 1.73} = \frac{4000}{397.9}$$

I = 10 AMPS

TO FIND HORSEPOWER

DIRECT CURRENT:

$$\text{HORSEPOWER} = \frac{\text{VOLTS} \times \text{AMPERES} \times \text{EFFICIENCY}}{746}$$

A 12 volt motor draws a current of 8.09 amperes, and has an efficiency rating of 96%. Determine the horsepower.

$$\text{HP} = \frac{E \times I \times \%EFF}{746} = \frac{12 \times 8.09 \times 0.96}{746} = \frac{93.19}{746}$$

HP = 0.1249 = 1/8 HP

SINGLE PHASE:

$$\text{HP} = \frac{\text{VOLTS} \times \text{AMPERES} \times \text{EFFICIENCY} \times \text{POWER FACTOR}}{746}$$

A single-phase, 115 volt (AC) motor has an efficiency rating of 92%, and a power factor of 80%. Determine the horsepower if the amp load is 4.4 amperes.

$$\text{HP} = \frac{E \times I \times \%EFF \times PF}{746} = \frac{115 \times 4.4 \times 0.92 \times 0.80}{746}$$

$$\text{HP} = \frac{372.416}{746} = 0.4992 = 1/2 \text{ HP}$$

THREE PHASE:

$$\text{HP} = \frac{\text{VOLTS} \times \text{AMPERES} \times \text{EFFICIENCY} \times \text{POWER FACTOR} \times 1.73}{746}$$

A three-phase, 460 volt motor draws a current of 52 amperes. The motor has an efficiency rating of 94%, and a power factor of 80%. Determine the horsepower.

$$\text{HP} = \frac{E \times I \times \%EFF \times PF \times 1.73}{746} = \frac{460 \times 52 \times 0.94 \times 0.80 \times 1.73}{746}$$

HP = 41.7 HP

TO FIND WATTS

The electrical power in any part of a circuit is equal to the current in that part multiplied by the voltage across that part of the circuit.

A watt is the power used when one volt causes one ampere to flow in a circuit.

One horsepower is the amount of energy required to lift 33,000 pounds, one foot, in one minute. The electrical equivalent of one horsepower is 745.6 watts. One watt is the amount of energy required to lift 44.26 pounds, one foot, in one minute. Watts is power, and power is the amount of work done in a given time.

When *VOLTS AND AMPERES* are known:

POWER (WATTS) = VOLTS x AMPERES

A 120 volt AC circuit draws a current of 5 amperes. Determine the power consumption.

P = E x I = 120 x 5 = 600 WATTS

We can now determine the resistance of this circuit.

POWER = RESISTANCE x (AMPERES)²

P = R x I² or 600 = R x 25 *divide both sides of equation by 25*

$\frac{600}{25}$ = R or R = 24 OHMS

or

POWER = $\frac{(VOLTS)^2}{RESISTANCE}$ or **P = $\frac{E^2}{R}$** or 600 = $\frac{120^2}{R}$

R x 600 = 120² or R = $\frac{14,400}{600}$ = 24 OHMS

NOTE: REFER TO THE FORMULAS OF THE OHM'S LAW CHART ON PAGE 24

TO FIND KILOWATTS

DIRECT CURRENT:

$$\text{KILOWATTS} = \frac{\text{VOLTS} \times \text{AMPERES}}{1000}$$

A 120 volt (DC) motor draws a current of 40 amperes. Determine the kilowatts.

$$\text{KW} = \frac{E \times I}{1000} = \frac{120 \times 40}{1000} = \frac{4800}{1000} = 4.8 \text{ KW}$$

SINGLE PHASE:

$$\text{KILOWATTS} = \frac{\text{VOLTS} \times \text{AMPERES} \times \text{POWER FACTOR}}{1000}$$

A single-phase, 115 volt (AC) motor draws a current of 20 amperes, and has a power-factor rating of 86%. Determine the kilowatts.

$$\text{KW} = \frac{E \times I \times PF}{1000} = \frac{115 \times 20 \times 0.86}{1000} = \frac{1978}{1000} = 1.978 = 2\text{KW}$$

THREE PHASE:

$$\text{KILOWATTS} = \frac{\text{VOLTS} \times \text{AMPERES} \times \text{POWER FACTOR} \times 1.73}{1000}$$

A three-phase, 460 volt (AC) motor draws a current of 52 amperes, and has a power-factor rating of 80%. Determine the kilowatts.

$$\text{KW} = \frac{E \times I \times PF \times 1.73}{1000} = \frac{460 \times 52 \times 0.80 \times 1.73}{1000}$$

$$= \frac{33,105}{1000} = 33.105 = 33\text{KW}$$

TO FIND KILOVOLT-AMPERES

SINGLE PHASE:

$$\text{KILOVOLT-AMPERES} = \frac{\text{VOLTS} \times \text{AMPERES}}{1000}$$

A single-phase, 240 volt generator delivers 41.66 amperes at full load. Determine the kilovolt-amperes rating.

$$\text{KVA} = \frac{E \times I}{1000} = \frac{240 \times 41.66}{1000} = \frac{10,000}{1000} = 10 \text{ KVA}$$

THREE PHASE:

$$\text{KILOVOLT-AMPERES} = \frac{\text{VOLTS} \times \text{AMPERES} \times 1.73}{1000}$$

A three-phase, 460 volt generator delivers 52 amperes. Determine the kilovolt-amperes rating.

$$\text{KVA} = \frac{E \times I \times 1.73}{1000} = \frac{460 \times 52 \times 1.73}{1000} = \frac{41,382}{1000}$$

$$= 41.382 = 41 \text{ KVA}$$

NOTE: KVA = APPARENT POWER = POWER BEFORE USED, SUCH AS THE RATING OF A TRANSFORMER.

Kirchhoff's Laws

FIRST LAW (CURRENT):
THE SUM OF THE CURRENTS ARRIVING AT ANY POINT IN A CIRCUIT MUST EQUAL THE SUM OF THE CURRENTS LEAVING THAT POINT.

SECOND LAW (VOLTAGE):
THE TOTAL VOLTAGE APPLIED TO ANY CLOSED CIRCUIT PATH IS ALWAYS EQUAL TO THE SUM OF THE VOLTAGE DROPS IN THAT PATH.
OR
THE ALGEBRAIC SUM OF ALL THE VOLTAGES ENCOUNTERED IN ANY LOOP EQUALS ZERO.

TO FIND INDUCTANCE

Inductance (L):

Inductance is the production of magnetization of electrification in a body by the proximity of a magnetic field or electric charge, or of the electric current in a conductor by the variation of the magnetic field in its vicinity. Expressed in Henrys.

A. To find the total inductance of coils connected in series.

$$L_T = L_1 + L_2 + L_3 + L_4$$

Determine the total inductance of four coils connected in series. Each coil has an inductance of four Henrys.

$$L_T = L_1 + L_2 + L_3 + L_4$$
$$= 4 + 4 + 4 + 4 = 16 \text{ Henrys}$$

B. To find the total inductance of coils connected in parallel.

$$\frac{1}{L_T} = \frac{1}{L_1} + \frac{1}{L_2} + \frac{1}{L_3} + \frac{1}{L_4}$$

Determine the total inductance of four coils connected in parallel. Each coil has an inductance of four Henrys.

$$\frac{1}{L_T} = \frac{1}{L_1} + \frac{1}{L_2} + \frac{1}{L_3} + \frac{1}{L_4}$$

$$\frac{1}{L_T} = \frac{1}{4} + \frac{1}{4} + \frac{1}{4} + \frac{1}{4}$$

$$\frac{1}{L_T} = \frac{4}{4} \text{ OR } L_T \times 4 = 1 \times 4 \text{ OR } L_T = \frac{4}{4} = 1 \text{ Henry}$$

An induction coil is a device, consisting of two concentric coils and an interrupter, that changes a low steady voltage into a high intermittent alternating voltage by electromagnetic induction. Most often used as a spark coil.

TO FIND IMPEDANCE

Impedance (Z):

Impedance is the total opposition to an alternating current presented by a circuit. Expressed in OHMS.

A. When *VOLTS AND AMPERES* are known:

$$\text{IMPEDANCE} = \frac{\text{VOLTS}}{\text{AMPERES}} \quad \text{OR} \quad Z = \frac{E}{I}$$

Determine the impedance of a 120 volt AC circuit that draws a current of four amperes.

$$Z = \frac{E}{I} = \frac{120}{4} = 30 \text{ OHMS}$$

B. When *RESISTANCE AND REACTANCE* are known:

$$Z = \sqrt{\text{RESISTANCE}^2 + \text{REACTANCE}^2} = \sqrt{R^2 + X^2}$$

Determine the impedance of an AC circuit when the resistance is 6 OHMS, and the reactance is 8 OHMS.

$$Z = \sqrt{R^2 + X^2} = \sqrt{36 + 64} = \sqrt{100} = 10 \text{ OHMS}$$

C. When *RESISTANCE, INDUCTIVE REACTANCE, AND CAPACITIVE REACTANCE* are known:

$$Z = \sqrt{R^2 + (X_L - X_C)^2}$$

Determine the impedance of an AC circuit that has a resistance of 6 OHMS, an inductive reactance of 18 OHMS, and a capacitive reactance of 10 OHMS.

$$Z = \sqrt{R^2 + (X_L - X_C)^2}$$
$$= \sqrt{6^2 + (18-10)^2} = \sqrt{6^2 + (8)^2}$$
$$= \sqrt{36 + 64} = \sqrt{100} = 10 \text{ OHMS}$$

TO FIND REACTANCE

Reactance (X):

Reactance in a circuit is the opposition to an alternating current caused by inductance and capacitance, equal to the difference between capacitive and inductive reactance. Expressed in OHMS.

A. INDUCTIVE REACTANCE X_L

Inductive reactance is that element of reactance in a circuit caused by self-inductance.

X_L = **2 x 3.1416 x FREQUENCY x INDUCTANCE**
   =   6.28   x   F   x   L

Determine the reactance of a four-Henry coil on a 60 cycle, AC circuit.

X_L = **6.28 x F x L** = 6.28 x 60 x 4 = 1507 OHMS

B. CAPACITIVE REACTANCE X_C

Capacitive reactance is that element of reactance in a circuit caused by capacitance.

$$X_C = \frac{1}{2 \times 3.1416 \times \text{FREQUENCY} \times \text{CAPACITANCE}}$$

$$= \frac{1}{6.28 \quad \times \quad F \quad \times \quad C}$$

Determine the reactance of a four microfarad condenser on a 60 cycle, AC circuit.

$$X_C = \frac{1}{6.28 \times F \times C} = \frac{1}{6.28 \times 60 \times .000004}$$

$$= \frac{1}{0.0015072} = 663 \text{ OHMS}$$

A HENRY is a unit of inductance, equal to the inductance of a circuit in which the variation of a current at the rate of one ampere per second induces an electromotive force of one volt.

SYNCHRONOUS SPEED

Synchronous Speed (in RPM) = $\dfrac{120f}{P}$

f = frequency
P = poles per phase

SLIP

Slip (in RPM) = Synchronous speed − actual speed

Slip percentage = $\dfrac{\text{Synchronous speed} - \text{actual speed}}{\text{Synchronous speed}} \times 100$

LOCKED-ROTOR CURRENT THREE PHASE

LRC = $\dfrac{1{,}000 \times HP \times kVA/HP}{V \times 1.732 \times PF \times eff}$

LRC = Locked rotor current, in amps
HP = Horsepower
V   = Volts
PF   = Power factor
eff = Motor efficiency

LOCKED-ROTOR CURRENT SINGLE PHASE

$$LRC = \frac{1{,}000 \times HP \times kVA/HP}{V \times 1.732 \times PF \times eff}$$

LRC = Locked rotor current, in amps
HP = Horsepower
V   = Volts
PF = Power factor
eff = Motor efficiency

MOTOR FRAMES

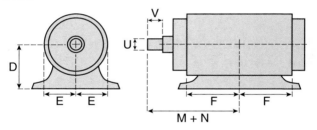

Motor frame dimensions are standardized, so that equipment and installation do not have to be customized around variable motor dimensions.

FRAME DIMENSIONS, IN INCHES

Frame No.	U	V	D	E	F	M & N	Keyway
42	3/8	–	2 5/8	1 3/4	27/32	4 1/32	–
48	1/2	–	2	2 1/8	1 3/8	5 3/8	–
56	5/8	–	3 1/2	2 7/16	1 1/2	6 1/8	3/16, 3/32
66	3/4	–	4 1/8	2 15/16	2 1/2	7 7/8	3/16, 3/32
143 T	7/8	2	3 1/2	2 3/4	2	6 1/2	3/16, 3/32
145 T	7/8	2	3 1/2	2 3/4	2 1/2	7	3/16, 3/32
182 T	1 1/8	2 1/2	4 1/2	3 3/4	2 1/4	7 3/4	1/4, 1/8
184 T	1 1/8	2 1/2	4 1/2	3 3/4	2 3/4	8 1/4	1/4, 1/8
213 T	1 3/8	3 1/8	5 1/4	4 1/4	2 3/4	9 5/8	5/16, 5/32
215 T	1 3/8	3 1/8	5 1/4	4 1/4	3 1/2	10 3/8	5/16, 5/32
254 T	1 5/8	3 3/4	6 1/4	5	4 1/8	12 3/8	3/8, 3/16
256 T	1 5/8	3 3/4	6 1/4	5	5	13 1/4	3/8, 3/16
284 T	1 7/8	4 3/8	7	5 1/2	4 3/4	14 1/8	1/2, 1/4
286 T	1 7/8	4 3/8	7	5 1/2	5 1/2	14 7/8	1/2, 1/4
324 T	2 1/8	5	8	6 1/4	5 1/4	15 3/4	1/2, 1/4
326 T	2 1/8	5	8	6 1/4	6	16 1/2	1/2, 1/4
364 T	2 3/8	5 5/8	9	7	5 5/8	17 3/8	5/8, 5/16
365 T	2 3/8	5 5/8	9	7	6 1/8	17 7/8	5/8, 5/16
404 T	2 7/8	7	10	8	6 1/8	20	3/4, 3/8
405 T	2 7/8	7	10	8	6 7/8	20 3/4	3/4, 3/8
444 T	3 3/8	8 1/4	11	9	7 1/4	23 1/4	7/8, 7/16
445 T	3 3/8	8 1/4	11	9	8 1/4	24 1/4	7/8, 7/16

FRAME LETTER DESIGNATIONS

G	Gasoline pump motor
K	Sump pump motor
M	Oil burner motor
N	Oil burner motor
S	Short shaft or direct connection
T	Standard dimensions
U	Old designation – standard dimensions
Y	Special dimensions required
Z	Shaft extension

INSULATION CLASSES

Class	°C
A	105
B	130
F	155
H	180

Class	°F
A	221
B	266
F	311
H	356

COMMON ELECTRICAL DISTRIBUTION SYSTEMS

120/240 Volt Single-Phase Three-Wire System

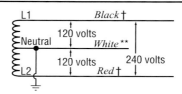

† • **Line one** ungrounded conductor colored **Black**.
† • **Line two** ungrounded conductor colored **Red**.
** • Grounded neutral conductor colored **White** or Gray.

120/240 Volt Three-Phase Four-Wire System (Delta High Leg)

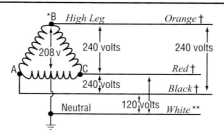

† • **A** phase ungrounded conductor colored **Black**.
†* • **B** phase ungrounded conductor colored **Orange** or tagged (High Leg). (Caution - 208V Orange to White)
† • **C** phase ungrounded conductor colored **Red**.
** • Grounded conductor colored **White** or Gray. (Center tap)

** Grounded conductors are required to be white or gray or three white stripes. See *NEC* 200.6A.

* B phase of high leg delta must be Orange or tagged.

† Ungrounded conductor colors may be other than shown; see local ordinances or specifications.

COMMON ELECTRICAL DISTRIBUTION SYSTEMS

120/208 Volt Three-Phase Four-Wire System (WYE Connected)

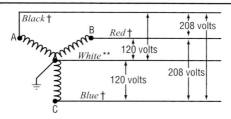

† • **A** phase ungrounded conductor colored **Black**.
† • **B** phase ungrounded conductor colored **Red**.
† • **C** phase ungrounded conductor colored **Blue**.
** • Grounded neutral conductor colored **White** or Gray.

277/480 Volt Three-Phase Four-Wire System (WYE Connected)

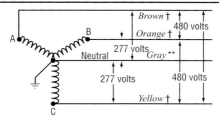

† • **A** phase ungrounded conductor colored **Brown**.
† • **B** phase ungrounded conductor colored **Orange**.
† • **C** phase ungrounded conductor colored **Yellow**.
** • Grounded neutral conductor colored **Gray**.

** Grounded conductors are required to be white or gray or three white stripes. See *NEC* 200.6A.

* B phase of high leg delta must be Orange or tagged.

† Ungrounded conductor colors may be other than shown; see local ordinances or specifications.

VOLTAGE DROP CALCULATIONS, INDUCTANCE NEGLIGIBLE

Vd = Voltage Drop
I = Current in Conductor (Amperes)
L = One-way Length of Circuit (Ft.)
Cm = Cross Section Area of Conductor (Circular Mils)
K = Resistance in ohms of one circular mil foot of conductor

K = 12.9 for Copper Conductors at 75°C
K = 21.2 for Aluminum Conductors at 75°C

NOTE: K value changes with temperature.
See *NEC* Chapter 9, Table 8, Notes

Single-Phase Circuits:

$$Vd = \frac{2K \times L \times I}{Cm} \quad \text{or} \quad {}^*Cm = \frac{2K \times L \times I}{Vd}$$

Three-Phase Circuits:

$$Vd = \frac{1.73K \times L \times I}{Cm} \quad \text{or} \quad {}^*Cm = \frac{1.73K \times L \times I}{Vd}$$

* NOTE: Always check ampacity tables to ensure conductors' ampacity is equal to load after voltage drop calculation.

VOLTAGE DROP CALCULATION EXAMPLES

DISTANCE (ONE-WAY) FOR 2% VOLTAGE DROP FOR 120 or 240 VOLTS - SINGLE PHASE
(60°C insulation & terminals)

AMPS	VOLTS	12 AWG	10 AWG	8 AWG	6 AWG	4 AWG	3 AWG	2 AWG	1	1/0 AWG
20	120	30	48	77	122	194	245	309	389	491
	240	60	96	154	244	388	490	618	778	982
30	120		32	51	81	129	163	206	260	327
	240		64	102	162	258	326	412	520	654
40	120			38	61	97	122	154	195	246
	240			76	122	194	244	308	390	492
50	120				49	78	98	123	156	196
	240				98	156	196	246	312	392
60	120					65	82	103	130	164
60	240					130	164	206	260	328
70	240					111	140	176	222	281
80	240						122	154	195	246
90	240							137	173	218
100	240								156	196

See footnotes on the next page concerning circuit load limitations.

VOLTAGE DROP CALCULATION EXAMPLES

Typical Voltage Drop Values Based on Conductor Size and One-Way Length* (60°C Termination and Insulation)

25 FEET

		12 AWG	10 AWG	8 AWG	6 AWG	4 AWG	3 AWG	2 AWG	1 AWG
AMPERES	20	1.98	1.24	0.78	0.49	0.31	0.25	0.19	0.15
	30		1.86	1.17	0.74	0.46	0.37	0.29	0.23
	40			1.56	0.98	0.62	0.49	0.39	0.31
	50				1.23	0.77	0.61	0.49	0.39
	60					0.93	0.74	0.58	0.46

50 FEET

		12 AWG	10 AWG	8 AWG	6 AWG	4 AWG	3 AWG	2 AWG	1 AWG
AMPERES	20	3.95	2.49	1.56	0.98	0.62	0.49	0.39	0.31
	30		3.73	2.34	1.47	0.93	0.74	0.58	0.46
	40			3.13	1.97	1.24	0.98	0.78	0.62
	50				2.46	1.55	1.23	0.97	0.77
	60					1.85	1.47	1.17	0.92

75 FEET

		12 AWG	10 AWG	8 AWG	6 AWG	4 AWG	3 AWG	2 AWG	1 AWG
AMPERES	20	5.93	3.73	2.34	1.47	0.93	0.74	0.58	0.46
	30		5.59	3.52	2.21	1.39	1.10	0.87	0.69
	40			4.69	2.95	1.85	1.47	1.17	0.92
	50				3.69	2.32	1.84	1.46	1.16
	60					2.78	2.21	1.75	1.39

100 FEET

		12 AWG	10 AWG	8 AWG	6 AWG	4 AWG	3 AWG	2 AWG	1 AWG
AMPERES	20	7.90	4.97	3.13	1.97	1.24	0.98	0.78	0.62
	30		7.46	4.69	2.95	1.85	1.47	1.17	0.92
	40			6.25	3.93	2.47	1.96	1.56	1.23
	50				4.92	3.09	2.45	1.94	1.54
	60					3.71	2.94	2.33	1.85

125 FEET

		12 AWG	10 AWG	8 AWG	6 AWG	4 AWG	3 AWG	2 AWG	1 AWG
AMPERES	20	9.88	6.21	3.91	2.46	1.55	1.23	0.97	0.77
	30		9.32	5.86	3.69	2.32	1.84	1.46	1.16
	40			7.81	4.92	3.09	2.45	1.94	1.54
	50				6.15	3.86	3.06	2.43	1.93
	60					4.64	3.68	2.92	2.31

150 FEET

		12 AWG	10 AWG	8 AWG	6 AWG	4 AWG	3 AWG	2 AWG	1 AWG
AMPERES	20	11.85	7.46	4.69	2.95	1.85	1.47	1.17	0.92
	30		11.18	7.03	4.42	2.78	2.21	1.75	1.39
	40			9.38	5.90	3.71	2.94	2.33	1.85
	50				7.37	4.64	3.68	2.92	2.31
	60					5.56	4.41	3.50	2.77

A two-wire 20 ampere circuit using 12 AWG with a one-way distance of 25 feet will drop 1.98 volts;
120 volts - 1.98 volts = 118.02 volts as the load voltage.
240 volts - 1.98 volts = 238.02 volts as the load voltage.
* Better economy and efficiency will result using the voltage drop method on page 44.
A continuous load cannot exceed 80% of the circuit rating.
A motor or heating load cannot exceed 80% of the circuit rating.

VOLTAGE DROP CALCULATION EXAMPLES

Single-Phase Voltage Drop

What is the voltage drop of a 240 volt single-phase circuit consisting of #8 THWN copper conductors feeding a 30 ampere load that is 150 feet in length?

Voltage Drop Formula

$$Vd = \frac{2K \times L \times I}{Cm} = \frac{2 \times 12.9 \times 150 \times 30}{16,510} = \frac{116,100}{16,510} = \mathbf{7} \text{ Volts}$$

Percentage voltage drop = 7 volts/240 volts = .029 = **2.9%**

Voltage at load = 240 volts - 7 volts = **233** volts

Three-Phase Voltage Drop

What is the voltage drop of a 480 volt three-phase circuit consisting of 250 kcmil THWN copper conductors that supply a 250 ampere load that is 500 feet from the source?

Voltage Drop Formula

250 kcmil = 250,000 circular mils

$$Vd = \frac{1.73K \times L \times I}{Cm} = \frac{1.73 \times 12.9 \times 500 \times 250}{250,000} = \frac{2,789,625}{250,000} = \mathbf{11} \text{ Volts}$$

Percentage voltage drop = 11 volts/480 volts = .0229 = **2.29%**

Voltage at load = 480 volts - 11 volts = **469** volts

NOTE: Always check ampacity tables for conductors selected.

MAXIMUM PERMISSIBLE CAPACITOR KVAR

For Use with Open-Type Three-Phase Sixty-Cycle Induction Motors

MOTOR RATING HP	3600 RPM		1800 RPM		1200 RPM	
	MAXIMUM CAPACITOR RATING KVAR	REDUCTION IN LINE CURRENT %	MAXIMUM CAPACITOR RATING KVAR	REDUCTION IN LINE CURRENT %	MAXIMUM CAPACITOR RATING KVAR	REDUCTION IN LINE CURRENT %
10	3	10	3	11	3.5	14
15	4	9	4	10	5	13
20	5	9	5	10	6.5	12
25	6	9	6	10	7.5	11
30	7	8	7	9	9	11
40	9	8	9	9	11	10
50	12	8	11	9	13	10
60	14	8	14	8	15	10
75	17	8	16	8	18	10
100	22	8	21	8	25	9
125	27	8	26	8	30	9
150	32.5	8	30	8	35	9
200	40	8	37.5	8	42.5	9

MOTOR RATING HP	900 RPM		720 RPM		600 RPM	
	MAXIMUM CAPACITOR RATING KVAR	REDUCTION IN LINE CURRENT %	MAXIMUM CAPACITOR RATING KVAR	REDUCTION IN LINE CURRENT %	MAXIMUM CAPACITOR RATING KVAR	REDUCTION IN LINE CURRENT %
10	5	21	6.5	27	7.5	31
15	6.5	18	8	23	9.5	27
20	7.5	16	9	21	12	25
25	9	15	11	20	14	23
30	10	14	12	18	16	22
40	12	13	15	16	20	20
50	15	12	19	15	24	19
60	18	11	22	15	27	19
75	21	10	26	14	32.5	18
100	27	10	32.5	13	40	17
125	32.5	10	40	13	47.5	16
150	37.5	10	47.5	12	52.5	15
200	47.5	10	60	12	65	14

NOTE: If capacitors of a lower rating than the values given in the table are used, the percentage reduction in line current given in the table should be reduced proportionately.

POWER-FACTOR CORRECTION

TABLE VALUES × KW OF CAPACITORS NEEDED TO CORRECT
FROM EXISTING TO DESIRED POWER FACTOR

EXISTING POWER FACTOR %	CORRECTED POWER FACTOR					
	100%	95%	90%	85%	80%	75%
50	1.732	1.403	1.247	1.112	0.982	0.850
52	1.643	1.314	1.158	1.023	0.893	0.761
54	1.558	1.229	1.073	0.938	0.808	0.676
55	1.518	1.189	1.033	0.898	0.768	0.636
56	1.479	1.150	0.994	0.859	0.729	0.597
58	1.404	1.075	0.919	0.784	0.654	0.522
60	1.333	1.004	0.848	0.713	0.583	0.451
62	1.265	0.936	0.780	0.645	0.515	0.383
64	1.201	0.872	0.716	0.581	0.451	0.319
65	1.168	0.839	0.683	0.548	0.418	0.286
66	1.139	0.810	0.654	0.519	0.389	0.257
68	1.078	0.749	0.593	0.458	0.328	0.196
70	1.020	0.691	0.535	0.400	0.270	0.138
72	0.964	0.635	0.479	0.344	0.214	0.082
74	0.909	0.580	0.424	0.289	0.159	0.027
75	0.882	0.553	0.397	0.262	0.132	
76	0.855	0.526	0.370	0.235	0.105	
78	0.802	0.473	0.317	0.182	0.052	
80	0.750	0.421	0.265	0.130		
82	0.698	0.369	0.213	0.078		
84	0.646	0.317	0.161			
85	0.620	0.291	0.135			
86	0.594	0.265	0.109			
88	0.540	0.211	0.055			
90	0.485	0.156				
92	0.426	0.097				
94	0.363	0.034				
95	0.329					

TYPICAL PROBLEM: With a load of 500 KW at 70% power factor, it is desired to find the KVA of capacitors required to correct the power factor to 85%

SOLUTION: From the table, select the multiplying factor 0.400 corresponding to the existing 70%, and the corrected 85% power factor. 0.400 x 500 = 200 KVA of capacitors required.

POWER FACTOR AND EFFICIENCY EXAMPLE

A squirrel-cage induction motor is rated 10 horsepower, 208 volt, three phase and has a nameplate rating of 27.79 amperes. A wattmeter reading indicates 8 kilowatts of consumed (true) power. Calculate apparent power (KVA), power factor, efficiency, internal losses, and size the capacitor in kilovolts reactive (KVAR) needed to correct the power factor to unity (100%).

Apparent input power: kilovolt-amperes (KVA)

KVA = (E × I × 1.73) / 1000 = (208 × 27.79 × 1.73) / 1000 = **10 KVA**

Power factor (PF) = ratio of true power (KW) to apparent power (KVA). Kilowatts / kilovolt-amperes = 8 KW/10 KVA = .8 = **80% Power Factor** 80% of the 10-KVA apparent power input performs work.

Motor output in kilowatts = 10 horsepower × 746 watts = 7460 watts = **7.46 KW.**

Efficiency = watts out/watts in = 7.46 KW / 8 KW = .9325 = **93.25% Efficiency.**

Internal losses (heat, friction, hysteresis) = 8 KW - 7.46 KW = **.54 KW** (540 watts)

Kilovolt-amperes reactive (KVAR) (Power stored in motor magnetic field)

$$KVAR = \sqrt{KVA^2 - KW^2} = \sqrt{10\ KVA^2 - 8\ KW^2} = \sqrt{100 - 64} = \sqrt{36}$$
= **6 KVAR**

The size capacitor needed to equal the motor's stored reactive power is 6 KVAR. (A capacitor stores reactive power in its electrostatic field).

POWER FACTOR AND EFFICIENCY EXAMPLE

The power source must supply the current to perform work and maintain the motor's magnetic field. Before power factor correction, this was 27.79 amperes. The motor magnetizing current after power factor correction is supplied by circulation of current between the motor and the electrostatic field of the capacitor and is no longer supplied by power source after initial start up.

The motor feeder current after correction to 100% will equal the amount required by the input watts in this case (8 KW × 1000) / (208 volts × 1.73) = **22.23 amps**

- Kilo = (1000 example: 1000 watts = 1 kilowatt)

- Inductive loads (motors, coils) have lagging currents and capacitive loads have leading currents.

- Inductance and capacitance have opposite effects in a circuit and can cancel each other out.

LOCKED-ROTOR CODE LETTERS

Letter Code	Kilovolt-Ampere per Horsepower with Locked Rotor	Letter Code	Kilovolt-Ampere per Horsepower with Locked Rotor
A	0 - 3.14	L	9.0 - 9.99
B	3.15 - 3.54	M	10.0 - 11.19
C	3.55 - 3.99	N	11.2 - 12.49
D	4.0 - 4.49	P	12.5 - 13.99
E	4.5 - 4.99	R	14.0 - 15.99
F	5.0 - 5.59	S	16.0 - 17.99
G	5.6 - 6.29	T	18.0 - 19.99
H	6.3 - 7.09	U	20.0 - 22.39
J	7.1 - 7.99	V	22.4 and up
K	8.0 - 8.99		

The *National Electrical Code*® requires that all alternating current motors rated 1/2 horsepower or more (except for polyphase wound-rotor motors) must have code letters on their nameplates indicating motor input with locked rotor (in kilovolt-amperes per horsepower). If you know the horsepower and voltage rating of a motor and its "Locked KVA per Horsepower" (from above table), you can calculate the locked-rotor current using the following formulas.

Adapted from *NEC*® Table 430.7(B).

LOCKED-ROTOR CODE LETTERS

Single-Phase Motors:

$$\text{Locked-Rotor Current} = \frac{HP \times KVA_{hp} \times 1000}{E}$$

Three-Phase Motors:

$$\text{Locked-Rotor Current} = \frac{HP \times KVA_{hp} \times 1000}{E \times 1.73}$$

Example: What is the maximum locked-rotor current for a 480 volt 25 horsepower code letter F motor?
(from the table on page 52, code letter F = 5.59 KVA_{hp})

$$I = \frac{HP \times KVA_{hp} \times 1000}{E \times 1.73} = \frac{25 \times 5.59 \times 1000}{480 \times 1.73} = \textbf{168.29 Amperes}$$

Adapted from *NEC*® Table 430.251(A).

THREE-PHASE AC MOTOR WINDINGS AND CONNECTIONS

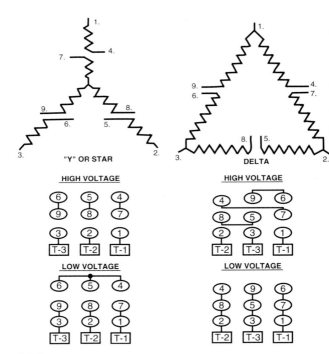

NOTES:
1. The most important part of any motor is the nameplate. Check the data given on the plate before making the connections.
2. To change rotation direction of three-phase motor, swap any 2 T-leads.

NEMA ENCLOSURE TYPES
NONHAZARDOUS LOCATIONS

The purpose of this document is to provide general information on the definitions of NEMA Enclosure Types to architects, engineers, installers, inspectors and other interested parties. [For more detailed and complete information, NEMA Standards Publication 250-2003, "Enclosures for Electrical Equipment (1000 Volts Maximum)" should be consulted.]

In **Nonhazardous Locations**, the specific enclosure types, their applications, and the environmental conditions they are designed to protect against, when completely and properly installed, are as follows:

Type 1 - Enclosures constructed for indoor use to provide a degree of protection to personnel against incidental contact with the enclosed equipment and to provide a degree of protection against falling dirt.

Type 2 - Enclosures constructed for indoor use to provide a degree of protection to personnel against incidental contact with the enclosed equipment, to provide a degree of protection against falling dirt, and to provide a degree of protection against dripping and light splashing of liquids.

Type 3 - Enclosures constructed for either indoor or outdoor use to provide a degree of protection to personnel against incidental contact with the enclosed equipment; to provide a degree of protection against falling dirt, rain, sleet, snow, and windblown dust; and that will be undamaged by the external formation of ice on the enclosure.

Type 3R - Enclosures constructed for either indoor or outdoor use to provide a degree of protection to personnel against incidental contact with the enclosed equipment; to provide a degree of protection against falling dirt, rain, sleet, and snow; and that will be undamaged by the external formation of ice on the enclosure.

Reprinted from NEMA 250-2003 by permission of the National Electrical Manufacturers Association.

NEMA ENCLOSURE TYPES NONHAZARDOUS LOCATIONS

Type 3S - Enclosures constructed for either indoor or outdoor use to provide a degree of protection to personnel against incidental contact with the enclosed equipment; to provide a degree of protection against falling dirt, rain, sleet, snow, and windblown dust; and in which the external mechanism(s) remain operable when ice laden.

Type 4 - Enclosures constructed for either indoor or outdoor use to provide a degree of protection to personnel against incidental contact with the enclosed equipment; to provide a degree of protection against falling dirt, rain, sleet, snow, windblown dust, splashing water, and hose-directed water; and that will be undamaged by the external formation of ice on the enclosure.

Type 4X - Enclosures constructed for either indoor or outdoor use to provide a degree of protection to personnel against incidental contact with the enclosed equipment; to provide a degree of protection against falling dirt, rain, sleet, snow, windblown dust, splashing water, hose-directed water, and corrosion; and that will be undamaged by the external formation of ice on the enclosure.

Type 5 - Enclosures constructed for indoor use to provide a degree of protection to personnel against incidental contact with the enclosed equipment; to provide a degree of protection against falling dirt; against settling airborne dust, lint, fibers, and flyings; and to provide a degree of protection against dripping and light splashing of liquids.

Type 6 - Enclosures constructed for either indoor or outdoor use to provide a degree of protection to personnel against incidental contact with the enclosed equipment; to provide a degree of protection against falling dirt; against hose-directed water and the entry of water during occasional temporary submersion at a limited depth; and that will be undamaged by the external formation of ice on the enclosure.

Reprinted from NEMA 250-2003 by permission of the National Electrical Manufacturers Association.

NEMA ENCLOSURE TYPES
NONHAZARDOUS LOCATIONS

Type 6P - Enclosures constructed for either indoor or outdoor use to provide a degree of protection to personnel against incidental contact with the enclosed equipment; to provide a degree of protection against falling dirt; against hose-directed water and the entry of water during prolonged submersion at a limited depth; and that will be undamaged by the external formation of ice on the enclosure.

Type 12 - Enclosures constructed (without knockouts) for indoor use to provide a degree of protection to personnel against incidental contact with the enclosed equipment; to provide a degree of protection against falling dirt; against circulating dust, lint, fibers, and flyings; and against dripping and light splashing of liquids.

Type 12K - Enclosures constructed (with knockouts) for indoor use to provide a degree of protection to personnel against incidental contact with the enclosed equipment; to provide a degree of protection against falling dirt; against circulating dust, lint, fibers, and flyings; and against dripping and light splashing of liquids.

Type 13 - Enclosures constructed for indoor use to provide a degree of protection to personnel against incidental contact with the enclosed equipment; to provide a degree of protection against falling dirt; against circulating dust, lint, fibers, and flyings; and against the spraying, splashing, and water, oil, and noncorrosive coolants.

Reprinted from NEMA 250-2003 *by permission of the National Electrical Manufacturers Association.*

NEMA ENCLOSURE TYPES HAZARDOUS LOCATIONS

In hazardous locations, when completely and properly installed and maintained, Type 7 and Type 10 enclosures are designed to contain an internal explosion without causing an external hazard. Type 8 enclosures are designed to prevent combustion through the use of oil-immersed equipment. Type 9 enclosures are designed to prevent the ignition of combustible dust.

Type 7 - Enclosures constructed for indoor use in hazardous locations classified as Class I, Division 1, Groups A, B, C, or D as defined in *NFPA 70*.

Type 8 - Enclosures constructed for either indoor or outdoor use in hazardous locations classified as Class I, Division 1, Groups A, B, C, or D as defined in *NFPA 70*.

Type 9 - Enclosures constructed for indoor use in hazardous conditions classified as Class II, Division 1, Groups E, F, or G as defined in *NFPA 70*.

Type 10 - Enclosures constructed to meet the requirements of the Mine Safety and Health Administration, 30 CFR, Part 18.

Reprinted from NEMA 250-2003 by permission of the National Electrical Manufacturers Association.

U.S. WEIGHTS AND MEASURES

Linear Measures

		1 INCH	= 2.540 CENTIMETERS
12 INCHES	= 1 FOOT		= 3.048 DECIMETERS
3 FEET	= 1 YARD		= 9.144 DECIMETERS
5.5 YARDS	= 1 ROD		= 5.029 METERS
40 RODS	= 1 FURLONG		= 2.018 HECTOMETERS
8 FURLONGS	= 1 MILE		= 1.609 KILOMETERS

Mile Measurements

1 STATUTE MILE	= 5,280 FEET
1 SCOTS MILE	= 5,952 FEET
1 IRISH MILE	= 6,720 FEET
1 RUSSIAN VERST	= 3,504 FEET
1 ITALIAN MILE	= 4,401 FEET
1 SPANISH MILE	= 15,084 FEET

Other Linear Measurements

1 HAND	= 4 INCHES	1 LINK	= 7.92 INCHES
1 SPAN	= 9 INCHES	1 FATHOM	= 6 FEET
1 CHAIN	= 22 YARDS	1 FURLONG	= 10 CHAINS
		1 CABLE	= 608 FEET

U.S. WEIGHTS AND MEASURES

Square Measures

144	SQUARE INCHES	= 1	SQUARE FOOT
9	SQUARE FEET	= 1	SQUARE YARD
30¼	SQUARE YARDS	= 1	SQUARE ROD
40	RODS	= 1	ROOD
4	ROODS	= 1	ACRE
640	ACRES	= 1	SQUARE MILE
1	SQUARE MILE	= 1	SECTION
36	SECTIONS	= 1	TOWNSHIP

Cubic or Solid Measures

1 CU. FOOT	=	1728	CU. INCHES
1 CU. YARD	=	27	CU. FEET
1 CU. FOOT	=	7.48	GALLONS
1 GALLON (WATER)	=	8.34	LBS.
1 GALLON (U.S.)	=	231	CU. INCHES OF WATER
1 GALLON (IMPERIAL)	=	277¼	CU. INCHES OF WATER

U.S. WEIGHTS AND MEASURES

Liquid Measurements

1 PINT   = 4 GILLS
1 QUART = 2 PINTS
1 GALLON = 4 QUARTS
1 FIRKIN = 9 GALLONS (ALE OR BEER)
1 BARREL = 42 GALLONS (PETROLEUM OR CRUDE OIL)

Dry Measures

1 QUART = 2 PINTS
1 PECK   = 8 QUARTS
1 BUSHEL = 4 PECKS

U.S. WEIGHTS AND MEASURES

Weight Measurements (Mass)

A. Avoirdupois Weight:

1 OUNCE	=	16 DRAMS
1 POUND	=	16 OUNCES
1 HUNDREDWEIGHT	=	100 POUNDS
1 TON	=	2000 POUNDS

B. Troy Weight:

1 CARAT	=	3.17 GRAINS
1 PENNYWEIGHT	=	20 GRAINS
1 OUNCE	=	20 PENNYWEIGHTS
1 POUND	=	12 OUNCES
1 LONG HUNDRED-WEIGHT	=	112 POUNDS
1 LONG TON	=	20 LONG HUNDREDWEIGHTS
	=	2240 POUNDS

C. Apothecaries Weight:

1 SCRUPLE	= 20 GRAINS	=	1.296 GRAMS
1 DRAM	= 3 SCRUPLES	=	3.888 GRAMS
1 OUNCE	= 8 DRAMS	=	31.1035 GRAMS
1 POUND	= 12 OUNCES	=	373.2420 GRAMS

D. Kitchen Weights and Measures:

1 U.S. PINT	=	16 FL. OUNCES
1 STANDARD CUP	=	8 FL. OUNCES
1 TABLESPOON	=	0.5 FL. OUNCES (15 CU. CMS.)
1 TEASPOON	=	0.16 FL. OUNCES (5 CU. CMS.)

METRIC SYSTEM

Prefixes:

A. MEGA = 1,000,000   E. DECI   = 0.1
B. KILO = 1000   F. CENTI = 0.01
C. HECTO = 100   G. MILLI = 0.001
D. DEKA = 10   H. MICRO = 0.000001

Linear Measures:

(THE UNIT IS THE METER = 39.37 INCHES)

1 CENTIMETER	= 10	MILLIMETERS	=	0.3937011 IN.
1 DECIMETER	= 10	CENTIMETERS	=	3.9370113 INS.
1 METER	= 10	DECIMETERS	=	1.0936143 YDS.
			=	3.2808429 FT.
1 DEKAMETER	= 10	METERS	=	10.936143 YDS.
1 HECTOMETER	= 10	DEKAMETERS	=	109.36143 YDS.
1 KILOMETER	= 10	HECTOMETERS	=	0.62137 MILE
1 MYRIAMETER	= 10,000 METERS			

Square Measures:

(THE UNIT IS THE SQUARE METER = 1549.9969 SQ. INCHES)

1 SQ. CENTIMETER	= 100 SQ. MILLIMETERS	= 0.1550 SQ. IN.
1 SQ. DECIMETER	= 100 SQ. CENTIMETERS	= 15.550 SQ. INS.
1 SQ. METER	= 100 SQ. DECIMETERS	= 10.7639 SQ. FT.
1 SQ. DEKAMETER	= 100 SQ. METERS	= 119.60 SQ. YDS.
1 SQ. HECTOMETER	= 100 SQ. DEKAMETERS	
1 SQ. KILOMETER	= 100 SQ. HECTOMETERS	

(THE UNIT IS THE "ARE" = 100 SQ. METERS)

1 CENTIARE	= 10	MILLIARES = 10.7643	SQ. FT.
1 DECIARE	= 10	CENTIARES = 11.96033	SQ. YDS.
1 ARE	= 10	DECIARES = 119.6033	SQ. YDS.
1 DEKARE	= 10	ARES = 0.247110	ACRES
1 HEKTARE	= 10	DEKARES = 2.471098	ACRES
1 SQ. KILOMETER	= 100	HEKTARES = 0.38611	SQ. MILE

METRIC SYSTEM

Cubic Measures:

(THE UNIT IS THE "STERE" = 61,025.38659 CU. INS.)
```
1 DECISTERE = 10 CENTISTERES =  3.531562 CU. FT.
1 STERE     = 10 DECISTERES  =  1.307986 CU. YDS.
1 DEKASTERE = 10 STERES      = 13.07986  CU. YDS.
```

(THE UNIT IS THE METER = 39.37 INCHES)
```
1 CU. CENTIMETER = 1000 CU. MILLIMETERS  =  0.06102 CU. IN.
1 CU. DECIMETER  = 1000 CU. CENTIMETERS  = 61.02374 CU. IN.
1 CU. METER      = 1000 CU. DECIMETERS   = 35.31467 CU. FT.
                 = 1 STERE               =  1.30795 CU. YDS.
1 CU. CENTIMETER (WATER)                 = 1 GRAM
1000 CU. CENTIMETERS (WATER) = 1 LITER   = 1 KILOGRAM
1 CU. METER (1000 LITERS)                = 1 METRIC TON
```

METRIC SYSTEM

Measures of Weight:

(THE UNIT IS THE GRAM = 0.035274 OUNCES)

1 MILLIGRAM	=	=	0.015432	GRAINS
1 CENTIGRAM	= 10 MILLIGRAMS	=	0.15432	GRAINS
1 DECIGRAM	= 10 CENTIGRAMS	=	1.5432	GRAINS
1 GRAM	= 10 DECIGRAMS	=	15.4323	GRAINS
1 DEKAGRAM	= 10 GRAMS	=	5.6438	DRAMS
1 HECTOGRAM	= 10 DEKAGRAMS	=	3.5274	OUNCES
1 KILOGRAM	= 10 HECTOGRAMS	=	2.2046223	POUNDS
1 MYRIAGRAM	= 10 KILOGRAMS	=	22.046223	POUNDS
1 QUINTAL	= 10 MYRIAGRAMS	=	1.986412	CWT.
1 METRIC TON	= 10 QUINTAL		= 2,2045.622	POUNDS

1 GRAM = 0.56438 DRAMS
1 DRAM = 1.77186 GRAMS
       = 27.3438 GRAINS
1 METRIC TON = 2,204.6223 POUNDS

 **METRIC SYSTEM**

Measures of Capacity:

(THE UNIT IS THE LITER = 1.0567 LIQUID QUARTS)

1 CENTILITER	= 10 MILLILITERS	=	0.338	FLUID OUNCES
1 DECILITER	= 10 CENTILITERS	=	3.38	FLUID OUNCES
1 LITER	= 10 DECILITERS	=	33.8	FLUID OUNCES
1 DEKALITER	= 10 LITERS	=	0.284	BUSHEL
1 HECTOLITER	= 10 DEKALITERS	=	2.84	BUSHELS
1 KILOLITER	= 10 HECTOLITERS	=	264.2	GALLONS

NOTE: $\dfrac{KILOMETERS}{8} \times 5 = MILES$ or $\dfrac{MILES}{5} \times 8 = KILOMETERS$

METRIC DESIGNATOR AND TRADE SIZES

METRIC DESIGNATOR												
12	16	21	27	35	41	53	63	78	91	103	129	155
³⁄₈	½	¾	1	1¼	1½	2	2½	3	3½	4	5	6
TRADE SIZE												

U.S. Weights and Measures/Metric Equivalent Chart

	In.	Ft.	Yd.	Mile	mm	cm	m	km
1 Inch =	1	.0833	.0278	1.578×10^{-5}	25.4	**2.54**	.0254	2.54×10^{-5}
1 Foot =	12	1	.333	1.894×10^{-4}	304.8	**30.48**	.3048	3.048×10^{-4}
1 Yard =	36	3	1	5.6818×10^{-4}	914.4	91.44	**.9144**	9.144×10^{-4}
1 Mile =	63,360	5280	1760	1	1,609,344	160,934.4	1,609.344	**1.609344**
1 mm =	**.03937**	.0032808	1.0936×10^{-3}	6.2137×10^{-7}	1	0.1	0.001	0.000001
1 cm =	**.3937**	.0328084	.0109361	6.2137×10^{-6}	10	1	0.01	0.00001
1 m =	39.37	3.28084	**1.09361**	6.2137×10^{-4}	1000	100	1	0.001
1 km =	39,370	3,280.84	1,093.61	**0.62137**	1,000,000	100,000	1000	1

In. = Inches   Ft. = Foot   Yd. = Yard   Mi. = Mile   mm = Millimeter   cm = Centimeter   m = Meter   km = Kilometer

Explanation of Scientific Notation:

Scientific Notation is simply a way of expressing very large or very small numbers in a more compact format. Any number can be expressed as a number between 1 and 10, multiplied by a power of 10 (which indicates the correct position of the decimal point in the original number). Numbers greater than 10 have positive powers of 10, and numbers less than 1 have negative powers of 10.

Example: $186,000 = 1.86 \times 10^5$ $0.000524 = 5.24 \times 10^{-4}$

METRIC DESIGNATOR AND TRADE SIZES

Useful Conversions/Equivalents

    1 BTU . . . . Raises 1 lb of water 1°F
    1 GRAM CALORIE . . . . Raises 1 gram of water 1°C
    1 CIRCULAR MIL . . . . . Equals 0.7854 sq. mil
    1 SQ. MIL . . . . Equals 1.27 cir. mils
    1 MIL . . . . Equals 0.001 in.

To determine circular mil of a conductor:

ROUND CONDUCTOR . . . . CM = (Diameter in mils)^2

BUS BAR . . . . CM = $\dfrac{\text{Width (mils)} \times \text{Thickness (mils)}}{0.7854}$

NOTES:
1 Millimeter = 39.37 Mils
1 Cir. Millimeter = 1550 Cir. Mils
1 Sq. Millimeter = 1974 Cir. Mils

DECIMAL EQUIVALENTS

FRACTION					DECIMAL
1/64					.0156
2/64	1/32				.0313
3/64					.0469
4/64	2/32	1/16			.0625
5/64					.0781
6/64	3/32				.0938
7/64					.1094
8/64	4/32	2/16	1/8		.125
9/64					.1406
10/64	5/32				.1563
11/64					.1719
12/64	6/32	3/16			.1875
13/64					.2031
14/64	7/32				.2188
15/64					.2344
16/64	8/32	4/16	2/8	1/4	.25
17/64					.2656
18/64	9/32				.2813
19/64					.2969
20/64	10/32	5/16			.3125
21/64					.3281
22/64	11/32				.3438
23/64					.3594
24/64	12/32	6/16	3/8		.375
25/64					.3906
26/64	13/32				.4063
27/64					.4219
28/64	14/32	7/16			.4375
29/64					.4531
30/64	15/32				.4688
31/64					.4844
32/64	16/32	8/16	4/8	2/4	.5

Decimals are rounded to the nearest 10,000th.

DECIMAL EQUIVALENTS

FRACTION					DECIMAL
33/64					.5156
34/64	17/32				.5313
35/64					.5469
36/64	18/32	9/16			.5625
37/64					.5781
38/64	19/32				.5938
39/64					.6094
40/64	20/32	10/16	5/8		.625
41/64					.6406
42/64	21/32				.6563
43/64					.6719
44/64	22/32	11/16			.6875
45/64					.7031
46/64	23/32				.7188
47/64					.7344
48/64	24/32	12/16	6/8	3/4	.75
49/64					.7656
50/64	25/32				.7813
51/64					.7969
52/64	26/32	13/16			.8125
53/64					.8281
54/64	27/32				.8438
55/64					.8594
56/64	28/32	14/16	7/8		.875
57/64					.8906
58/64	29/32				.9063
59/64					.9219
60/64	30/32	15/16			.9375
61/64					.9531
62/64	31/32				.9688
63/64					.9844
64/64	32/32	16/16	8/8	4/4	1.000

Decimals are rounded to the nearest 10,000th.

SINGLE-PHASE MOTORS

SPLIT-PHASE—SQUIRREL-CAGE—DUAL-VOLTAGE

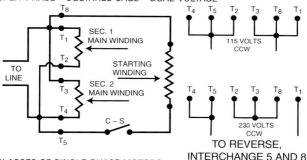

CLASSES OF SINGLE-PHASE MOTORS:
1. SPLIT-PHASE
    A. CAPACITOR-START
    B. REPULSION-START
    C. RESISTANCE-START
    D. SPLIT-CAPACITOR

2. COMMUTATOR
    A. REPULSION
    B. SERIES

TERMINAL COLOR MARKING:

T_1 <u>BLUE</u>	T_3 <u>ORANGE</u>	T_5 <u>BLACK</u>
T_2 <u>WHITE</u>	T_4 <u>YELLOW</u>	T_8 <u>RED</u>

NOTE: Split-phase motors are usually fractional horsepower. The majority of electric motors used in washing machines, refrigerators, etc. are of the split-phase type.

To change the speed of a split-phase motor, the number of poles must be changed.

1. Addition of running winding
2. Two starting windings, and two running windings
3. Consequent pole connections

SINGLE-PHASE MOTORS

SPLIT-PHASE—SQUIRREL-CAGE

A. RESISTANCE START:

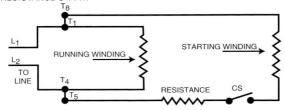

Centrifugal switch (CS) opens after reaching 75% of normal speed.

B. CAPACITOR START:

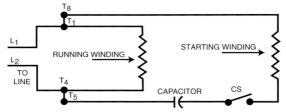

NOTES:
1. A resistance start motor has a resistance connected in series with the starting winding.
2. The capacitor start motor is employed where a high starting torque is required.

DIRECT-CURRENT MOTORS

TERMINAL MARKINGS:
Terminal markings are used to tag terminals to which connections are to be made from outside circuits.

Facing the end opposite the drive (commutator end) the standard direction of shaft rotation is counter-clockwise.

- A-1 and A-2 indicate armature leads.
- S-1 and S-2 indicate series-field leads.
- F-1 and F-2 indicate shunt-field leads.

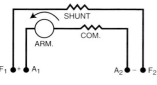

SHUNT-WOUND MOTORS
To change rotation, reverse either armature leads or shunt leads. <u>Do not</u> reverse both armature and shunt leads.

SERIES-WOUND MOTORS
To change rotation, reverse either armature leads or series leads. <u>Do not</u> reverse both armature and series leads.

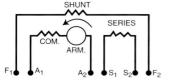

COMPOUND-WOUND MOTORS
To change rotation, reverse either armature leads or both the series and shunt leads. <u>Do not</u> reverse all three sets of leads.

NOTE: Standard rotation for <u>DC generator</u> is clockwise.

MOTOR SELECTION CHECKLIST

1. Horsepower requirements
2. Torque requirements
3. Speed requirements
4. Position (vertical, horizontal, etc.)
5. Conditions (temperature, water, corrosion, dust, etc.)
6. Operating cycle
7. Direction of rotation
8. Endplay
9. Available voltage, phases, frequency
10. Available starting current
11. Power factor concerns

MOTOR SELECTION CRITERIA

1. <u>Enclosures</u>. Must be suitable to area of installation, but enclosed motors are more expensive for both purchase and operation.

2. <u>Torque</u>. Motor's torque must exceed maximum torque requirements of the load. High-slip motors are preferred where there will be frequent peaks in required torque.

3. <u>Load cycle</u>. If loads cycle regularly, an average load (by RMS method) may be assumed, with a safety margin.

4. <u>Loading</u>. Motors operate most efficiently when fully loaded.

5. <u>Ambient temperatures</u>. Motors are designed for an ambient operating temperature of 40°C. Each 10° above this will halve the life of types A & B winding insulation.

TYPICAL LOAD SERVICE FACTORS

Load	Service Factor
Pump – centrifugal	1.0
Pump – centrifugal, sewage	2.0
Pump – rotary	1.5
Pump – reciprocating	2.0
Fan – light-duty	1.0
Fan – centrifugal	1.5
Blower – centrifugal	1.0
Blower – vane	1.25
Compressor – centrifugal	1.25
Compressor – vane	1.5
Elevator – bucket	2.0
Elevator – freight	2.25
Conveyor	1.5
Conveyor – heavy use	2.0
Punch press	2.25
Extruder – plastic	2.0
Extruder – metal	2.5
Concrete mixer	2.0
Printing press	1.5
Woodworking machines	1.0

OVERCURRENT PROTECTION FOR TWO OR MORE MOTORS

1. Determine the size of an overcurrent protection device, sized for the largest motor.

2. Add size determined in step 1 to the full-load amperage of all other motors.

3. The group overcurrent device for the motor feeder may be no larger than the amperage arrived at in step 2.

See *NEC* sections:

- 430.62(A) on overcurrent devices
- 430.24 on feeders
- 366.22 on gutters (if necessary)

OVERCURRENT PROTECTION FOR MOTORS AND OTHER LOADS

1. Determine the size of an overcurrent protection device, based upon the largest motor.

2. Add the sum of all other loads to size determined in step 1.

3. The group overcurrent device may be no larger than the amperage arrived at in step 2.

See *NEC* sections:

- 430.62(A) on overcurrent devices
- 215.2(A)(1) on feeders
- 250.142(B) on sub-panels
- 240.6(A) on overcurrent device sizes

DETERMINING OVERLOAD SIZE

1. Find motor full-load amps.

2. Add 25% (to achieve 125% total).

3. Choose overloads based upon step 2.

Example: If full-load current is 60 amps, overloads must be chosen based upon 60 × 1.25, or 75 amps.

See *NEC* sections:

- 430.6(A)(2) on full-load currents
- 430.32(A)(1) on adding 25%

DETERMINING CONTROLLER SIZE

1. Determine horsepower size of motor.
2. Controller may have a horsepower rating no less than horsepower size of motor.

See *NEC* sections:

- 430.83(A)(1) on controllers
- 430.110(A) on disconnecting means

MOTORS 2 HORSEPOWER OR LESS AND 300 VOLTS OR LESS

Switches may be used as controllers provided:

1. A general use switch can be used if it is rated for at least twice the full-load current of the motor.
2. AC switches on AC circuits may be used if the switch is rated 125% of the full-load current of the motor.

See *NEC* sections:

- 430.83(C)(1) on general use switches
- 430.83(C)(2) on AC switches

DETERMINING CONDUCTOR SIZES FOR SINGLE-PHASE MOTORS

1. Find motor full-load amps.
2. Add 25% to full-load amps (to achieve 125%).
3. Conductor must have an ampacity no lower than determined in step 2.
4. Select conductors from *NEC* Table 310.16.

See *NEC* sections:

- 430.6(A)(1) on full-load currents
- 430.22(A) on adding 25%
- 310.10, FPN(2) on conductors

DETERMINING CONDUCTOR SIZES FOR ADJUSTABLE SPEED DRIVES

1. Determine rated input of power conversion equipment.
2. Add 25% (to achieve 125%) of rated current.
3. Select conductors from *NEC* Table 310.16.

See *NEC* sections:

- 430.122(A) on adjustable speed drive conductors
- 430.2 on variable speed motors

MOTOR AND MOTOR-CIRCUIT CONDUCTOR PROTECTION

Motors can have large starting currents three to five times or more than that of the actual motor current. In order for motors to start, the motor and motor circuit conductors are allowed to be protected by circuit breakers and fuses at values that are higher than the actual motor and conductor ampere ratings. These larger overcurrent devices do not provide overload protection and will only open upon short circuits or ground faults. Overload protection must be used to protect the motor based on the actual nameplate amperes of the motor. This protection is usually in the form of heating elements in manual or magnetic motor starters. Small motors such as waste disposal motors have a red overload reset button built into the motor.

 GENERAL MOTOR RULES

- Use Full-Load Current from tables instead of nameplate.
- Branch-Circuit Conductors - Use 125% of Full-Load Current to find conductor size.
- Branch-Circuit OCP Size - Use percentages given in tables for Full-Load Current.
- Feeder Conductor Size - 125% of largest motor and sum of the rest.
- Feeder OCP - Use largest OCP plus rest of Full-Load Currents.

MOTOR BRANCH CIRCUIT AND FEEDER EXAMPLE

General Motor Applications

Branch-Circuit Conductors: Use Full-Load Three-Phase Currents;
From *NEC* Table 430.250,
50 HP 480 Volt Three-Phase motor design B, 75 degree terminations
= 65 Amperes
125% of Full-Load Current [*NEC* 430.22(A)]
125% of 65 A = **81.25 Amperes** Conductor Selection Ampacity

Branch-Circuit Overcurrent Device: *NEC* 430.52 (C1)
(Branch-Circuit Short Circuit and Ground-Fault Protection)
Use percentages given in 2005 *NEC* 430.52 for **Type**
of circuit breaker or fuse used.
50 HP 480 V 3 Ph Motor = 65 Amperes.
Nontime Fuse = 300%.
300% of 65A = 195 A. *NEC* 430.52(C1)(EX1) Next size allowed
NEC 240. 6A = **200 Ampere Fuse**.

MOTOR BRANCH CIRCUIT AND FEEDER EXAMPLE

Feeder Connectors: For 50 HP and 30 HP 480 Volt Three-Phase design B motors on same feeder
Use 125% of largest full-load current and 100% of rest. (*NEC* 430.24)
50 HP 480 V 3 Ph Motor = 65A; 30 HP 480 V 3 Ph Motor = 40A
(125% of 65A) + 40A = **121.25 A** Conductor Selection Ampacity

Feeder Overcurrent Device: [*NEC* 430.62(A)]
(Feeder short circuit and ground-fault protection)
Use largest overcurrent protection device <u>plus</u> full-load currents of the rest of the motors.
50 HP = 200 A fuse   (65 FLC)
30 HP = 125 A fuse   (40 FLC)
200 A fuse + 40 A (FLC) = 240 A. Do not exceed this value on feeder. Go down to a **225 A** fuse.

APPROXIMATE TORQUE FIGURES, COMPOUND DC MOTORS

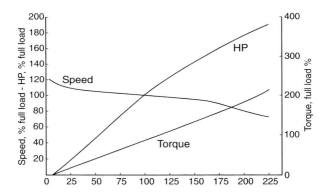

Speed and torque are inversely proportional; as one rises, the other falls.

HEATER CORRECTIONS FOR AMBIENT TEMPERATURES

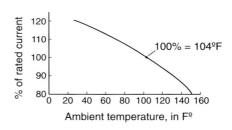

100% is standardized at 104°F (40°C). Electrical components must be protected from excess heat.

APPROXIMATE TORQUE FIGURES, AC MOTORS

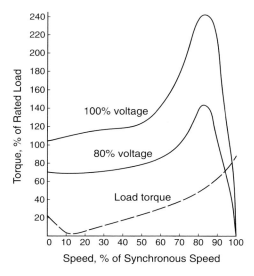

A standard AC motor may never reach synchronous speed. As it approaches synchronous speed, torque falls to zero and rotation slows.

APPROXIMATE TORQUE FIGURES, WOUND ROTOR MOTORS

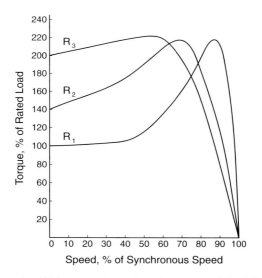

Torque peaks at higher percentages of synchronous speed, but falls to zero as synchronous speed is approached.

SIZING LOAD CONDUCTORS FROM GENERATORS

1. Determine generator output current from nameplate.
2. Add 15% (to achieve 115%) to generator current.
3. Select conductors from *NEC* Table 310.16.

See *NEC* sections:

- 445.13 on conductor ampacity
- 445.13, exception, on an exception to the 15% adder
- 310.10, FPN(2) on conductors

ELECTRICAL SAFETY DEFINITIONS

Arc Flash - The sudden release of heat energy and intense light at the point of an arc. Can be considered a short circuit through the air, usually created by accidental contact between live conductors.

Arc Blast - A pressure wave created by the heating, melting, vaporization, and expansion of conducting material and surrounding gases of air.

Arc Gap - The distance between energized conductors or between energized conductors and ground. Shorter arc gaps result in less energy being expended in the arc, while longer gaps reduce arc current. For 600 volts and below, arc gaps of 1.25 inches (32 mm) typically produce the maximum incident energy.

Approach Boundaries - Protection boundaries established to protect personnel from shock.

Calorie - The amount of heat needed to raise the temperature of 1 gram of water by 1° Celsius. 1 cal/cm^2 is equivalent to the exposure on the tip of a finger by a cigarette lighter for 1 second.

Distance to Arc - Refers to the distance from the receiving surface to the arc center. The value used for most calculations is typically 18 inches.

Electrically Safe Work Condition - Condition where the equipment and or circuit components have been disconnected from electrical energy sources, locked/tagged out, and tested to verify all sources of power are removed.

Reprinted with permission from Littelfuse®; www.littelfuse.com; 1-800-TEC-FUSE
For more information, refer to *NFPA 70E, Standard for Electrical Safety in the Workplace.*

ELECTRICAL SAFETY DEFINITIONS

Exposed Live Parts - An energized conductor or part that is capable of being inadvertently touched or approached (nearer than a safe distance) by a person. It is applicable to parts that are not in an electrically safe work condition, suitably grounded, isolated, or insulated.

Flame Resistant (FR) - A term referring to fabric and its ability to limit the ignition or burning of the garment. It can be a specific characteristic of the material or a treatment applied to a material.

Flash Hazard Analysis - A study that analyzes potential exposure to Arc-Flash hazards. The outcome of the study establishes Incident Energy levels, Hazard Risk Categories, Flash Protection Boundaries, and required PPE. It also helps define safe work practices.

Flash Protection Boundary - A protection boundary established to protect personnel from Arc-Flash hazards. The Flash Protection Boundary is the distance at which an unprotected worker can receive a second-degree burn to bare skin.

Flash Suit - A term referring to a complete FR rated Personal Protective Equipment (PPE) system that would cover a person's body, excluding the hands and feet. Included would be pants, shirt/jacket, and flash hood with a built-in face shield.

Hazard Risk Category - A classification of risks (from 0–4) defined by *NFPA 70E*. Each category requires PPE and is related to incident energy levels.

Reprinted with permission from Littelfuse®; www.littelfuse.com; 1-800-TEC-FUSE
For more information, refer to *NFPA 70E, Standard for Electrical Safety in the Workplace*.

ELECTRICAL SAFETY DEFINITIONS

Incident Energy - The amount of thermal energy impressed on a surface generated during an electrical arc at a certain distance from the arc. Typically measured in cal/cm^2.

PPE - An acronym for Personal Protective Equipment. It can include clothing, tools, and equipment.

Qualified Person - A person who is trained and knowledgeable on the construction and operation of the equipment and can recognize and avoid electrical hazards that may be encountered.

Shock - A trauma subjected to the body by electrical current. When personnel come in contact with energized conductors, it can result in current flowing through their body, often causing serious injury or death.

Unqualified Person - A person who does not possess all the skills and knowledge or has not been trained for a particular task.

Reprinted with permission from Littelfuse®; www.littelfuse.com; 1-800-TEC-FUSE

For more information, refer to *NFPA 70E, Standard for Electrical Safety in the Workplace.*

ELECTRICAL SAFETY CHECKLIST

1. Deenergize the equipment whenever possible prior to performing any work.

2. Verify you are "qualified" and properly trained to perform the required task.

3. Identify the equipment and verify you have a clear understanding and have been trained on how the equipment operates.

4. Provide justification why the work must be performed in an "energized" condition (if applicable).

5. Identify which safe work practices will be used to insure safety.

6. Determine if a Hazard Analysis has been performed to identify all hazards (Shock, Arc Flash, etc.).

7. Identify protection boundaries for Shock (Limited, Restricted & Prohibited Approach) and Arc Flash (Flash Protection Boundary).

8. Identify the required Personal Protective Equipment (PPE) for the task to be performed based on the Hazard Risk Category (HRC) and available incident Energy (cal/cm^2).

9. Provide barriers or other means to prevent access to the work area by "unqualified" workers.

10. Perform a job briefing and identify job-or task-specific hazards.

11. Obtain written management approval to perform the work in an "energized" condition (where applicable).

Reprinted with permission from Littelfuse®; www.littelfuse.com; 1-800-TEC-FUSE
For more information, refer to *NFPA 70E, Standard for Electrical Safety in the Workplace.*

ELECTRICAL SAFETY LOCKOUT–TAGOUT PROCEDURES

OSHA requires that energy sources to machines or equipment must be turned off and disconnected isolating them from the energy source. The isolating or disconnecting means must be either locked or tagged with a warning label. While lockout is the more reliable and preferred method, OSHA accepts tagout to be a suitable replacement in limited situations. *NFPA 70E* Article 120 contains detailed instructions for lockout–tagout and placing equipment in an electrically safe work condition.

Application of Lockout–Tagout Devices

1. Make necessary preparations for shutdown.
2. Shut down the machine or equipment.
3. Turn OFF (open) the energy isolating device (fuse/circuit breaker).
4. Apply the lockout or tagout device.
5. Render safe all stored or residual energy.
6. Verify the isolation and deenergization of the machine or equipment.

Reprinted with permission from Littelfuse®; www.littelfuse.com; 1-800-TEC-FUSE
For more information, refer to *NFPA 70E, Standard for Electrical Safety in the Workplace.*

ELECTRICAL SAFETY LOCKOUT–TAGOUT PROCEDURES

Removal of Lockout–Tagout Devices

1. Inspect the work area to ensure that nonessential items have been removed and that machine or equipment components are intact and capable of operating properly. Especially look for tools or pieces of conductors that may have not been removed.

2. Check the area around the machine or equipment to ensure that all employees have been safely positioned or removed.

3. Make sure that only the employees who attached the locks or tags are the ones removing them.

4. After removing locks or tags, notify affected employees before starting equipment or machines.

NOTE: For specific lockout–tagout procedures, refer to OSHA and NFPA 70E.

Reprinted with permission from Littelfuse®; www.littelfuse.com; 1-800-TEC-FUSE
For more information, refer to *NFPA 70E, Standard for Electrical Safety in the Workplace.*

ELECTRICAL SAFETY SHOCK PROTECTION BOUNDARIES

Nominal System Voltage (Phase to Phase)	Limited Approach Boundary		Restricted Approach Boundary	Prohibited Approach Boundary
	Exposed Fixed Circuit Part	Exposed Movable Conductor		
50 to 300 V	10 ft. 0 in.	3 ft. 6 in.	Avoid Contact	Avoid Contact
301 to 750 V	10 ft. 0 in.	3 ft. 6 in.	1 ft. 0 in.	0 ft. 1 in.
751 V to 15 kV	10 ft. 0 in.	5 ft. 0 in.	2 ft. 2 in.	0 ft. 7 in.
15.1 kV to 36 kV	10 ft. 0 in.	6 ft. 0 in.	2 ft. 7 in.	0 ft. 10 in.
36.1 kV to 46 kV	10 ft. 0 in.	8 ft. 0 in.	2 ft. 9 in.	1 ft. 5 in.
46.1 kV to 72.5 kV	10 ft. 0 in.	8 ft. 0 in.	3 ft. 2 in.	2 ft. 1 in.
72.6 kV to 121 kV	10 ft. 8 in.	8 ft. 0 in.	3 ft. 3 in.	2 ft. 8 in.

NOTE: Data derived from NFPA 70E Table 130.2(C)

Shock protection boundaries are based on system voltage and whether the exposed energized components are fixed or movable. *NFPA 70E Table 130.2(C)* defines these boundary distances for nominal phase-to-phase system voltages from 50 Volts to 800 kV. Approach Boundary distances may range from an inch to several feet. Please refer to *NFPA 70E Table 130.2(C)* for more information.

Protection Boundaries:

Limited Approach: Qualified person or unqualified person if accompanied by qualified person. PPE is required.

Restricted Approach: Qualified persons only. PPE is required.

Prohibited Approach: Qualified persons only. Use PPE as if making direct contact with a live part.

Reprinted with permission from Littelfuse®; www.littelfuse.com; 1-800-TEC-FUSE

For more information, refer to *NFPA 70E, Standard for Electrical Safety in the Workplace.*

ELECTRICAL SAFETY
HOW TO READ A WARNING LABEL

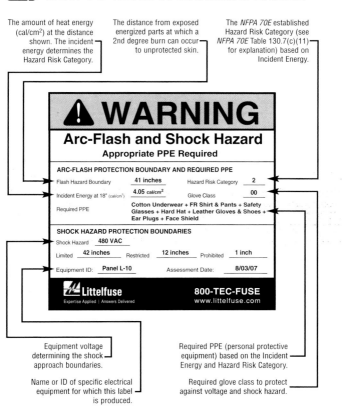

Reprinted with permission from Littelfuse®; www.littelfuse.com; 1-800-TEC-FUSE

PULLEY CALCULATIONS

The most common configuration consists of a motor with a pulley attached to its shaft, connected by a belt to a second pulley. The motor pulley is referred to as the **Driving Pulley**. The second pulley is called the **Driven Pulley**. The speed at which the Driven Pulley turns is determined by the speed at which the Driving Pulley turns as well as the diameters of both pulleys. The following formulas may be used to determine the relationships between the motor, pulley diameters and pulley speeds.

D = Diameter of Driving Pulley
d^1 = Diameter of Driven Pulley
S = Speed of Driving Pulley (revolutions per minute)
s^1 = Speed of Driven Pulley (revolutions per minute)

- *To determine the speed of the Driven Pulley (Driven RPM):*

  $$s^1 = \frac{D \times S}{d^1} \quad \text{or} \quad \text{Driven RPM} = \frac{\text{Driving Pulley Dia.} \times \text{Driving RPM}}{\text{Driven Pulley Dia.}}$$

- *To determine the speed of the Driving Pulley (Driving RPM):*

  $$S = \frac{d^1 \times s^1}{D} \quad \text{or} \quad \text{Driving RPM} = \frac{\text{Driven Pulley Dia.} \times \text{Driven RPM}}{\text{Driving Pulley Dia.}}$$

- *To determine the diameter of the Driven Pulley (Driven Dia.):*

  $$d^1 = \frac{D \times S}{s^1} \quad \text{or} \quad \text{Driven Dia.} = \frac{\text{Driving Pulley Dia.} \times \text{Driving RPM}}{\text{Driven RPM}}$$

- *To determine the diameter of the Driving Pulley (Driving Dia.):*

  $$D = \frac{d^1 \times s^1}{S} \quad \text{or} \quad \text{Driving Dia.} = \frac{\text{Driven Pulley Dia.} \times \text{Driven RPM}}{\text{Driving RPM}}$$

DETERMINING BELT LENGTH

Length $= \dfrac{\pi(D+d)}{2} + \sqrt{x^2 + (\dfrac{D-d}{2})^2}$

D = Diameter of larger pulley
d = Diameter of smaller pulley
π = 3.1416
x = Distance between shaft centers

HORSEPOWER CAPACITIES

Belt Speed (ft./min)	Pulley Diameter						
	1/2"	1"	1½"	2"	3"	4"	6"
1000	.33	.53	.66	.79	.92	.99	1.05
2000	.99	1.32	1.38	1.45	1.65	1.78	1.91
3000	1.25	1.58	1.98	2.17	2.59	2.70	2.90
4000	1.51	1.91	2.44	2.77	3.17	3.43	3.76
5000	1.65	2.11	2.77	3.10	3.63	3.96	4.42
6000	1.71	2.31	3.03	3.36	3.96	4.42	4.95
7000	1.71	2.37	3.16	3.56	4.29	4.88	5.54
8000	1.65	2.44	3.30	3.69	4.49	4.55	5.87
9000	1.51	2.50	3.36	3.76	4.55	4.62	6.07
10,000	1.32	2.50	3.43	3.82	4.62	4.62	6.14

NOTES:
Based on medium, Single-Ply, Dacron Belts
per inch of width

GEAR SIZING

$N = \dfrac{n \times R}{r}$

$n = \dfrac{N \times R}{r}$

n = Number of teeth, driven gear
N = Number of teeth, driving gear
R = RPM, pinion
r = RPM, gear

 DETERMINING SHAFT DIAMETER

Shaft diameter (in inches) = $\sqrt{\dfrac{K \times HP}{RPM}}$

HP = Horsepower transmitted
K = Constant, varying between 50 and 125, depending on shaft and distance between bearings

GEAR REDUCERS

Output Torque

$O_T = I_T \times R_R \times R_E$

Output Speed

$O_S = \dfrac{I_S}{R_R} \times R_E$

Output Horsepower

$O_{HP} = I_{HP} \times R_E$

O_T = Output torque
I_T = Input torque
R_R = Gear reduction ratio
O_S = Output speed (RPM)
I_S = Input speed (RPM)
R_E = Reducer efficiency
O_{HP} = Output horsepower
I_{HP} = Input horsepower

MOTOR TORQUE

Torque

$$T = \frac{HP \times 5252}{RPM}$$

Starting Torque

$$T = \frac{HP \times 5252 \times C}{RPM}$$

T = Torque, 1 lb-ft
HP = Horsepower
RPM= Rotations per minute
C = Motor class percentage

5252 is a constant, derived as follows: $\frac{33,000 \; 1 \, lb\text{-}ft}{\pi \times 2} = 5252$

CALCULATING COST OF OPERATING AN ELECTRICAL APPLIANCE

What is the monthly cost of operating a 240 volt 5 kilowatt (kW) central electric heater that operates 12 hours per day, when the cost is 15 cents per kilowatt-hour (kWhr)?

Cost = Watts x Hours used x Rate per kWhr / 1000

5 kW = 5000 Watts
Hours = 12 hours x 30 days = 360 hours per month

= 5000 x 360 x .15 / 1000
= 270,000 / 1000 = **$270 Monthly cost**

The above example is for a resistive load. Air-conditioning loads are primarily inductive loads. However, if ampere and voltage values are known this method will give an approximate cost. Kilowatt-hour rates vary for different power companies, and for residential use, graduated rate scales are usually used (the more power used, the lower the rate). Commercial and industrial rates are generally based on kilowatt usage, maximum demand, and power factor. Other costs are often added such as fuel cost adjustments.

ELECTRICAL SYMBOLS

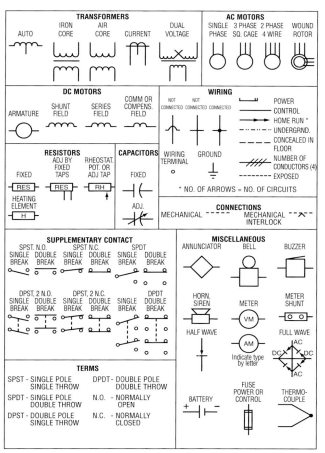

ELECTRICAL SYMBOLS

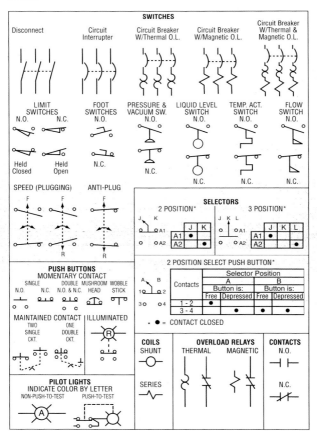

WIRING DIAGRAMS

Basic Diagram of Two-Wire Control Circuit

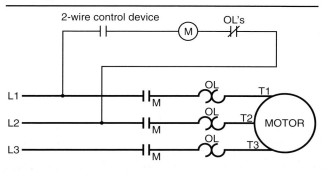

Wiring Diagram of Starter (Two-Wire Control)

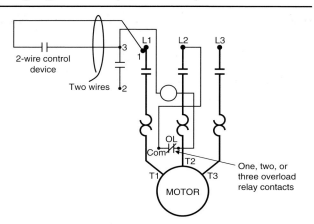

Reprinted from Miller, Charles R. *NFPA's Pocket Electrical References, First Edition.* Jones and Bartlett Publishers, 2007.

WIRING DIAGRAMS

Control Circuit Only

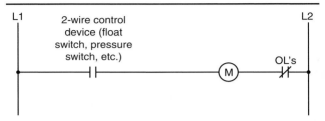

Basic Three-Wire Control Circuit

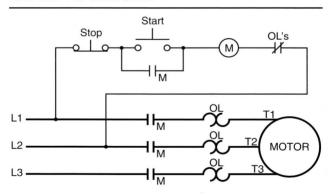

Reprinted from Miller, Charles R. *NFPA's Pocket Electrical References, First Edition.* Jones and Bartlett Publishers, 2007.

WIRING DIAGRAMS

Wiring Diagram of Starter (Three-Wire Control)

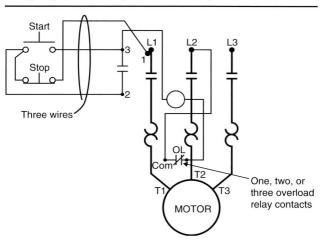

Control Circuit Only

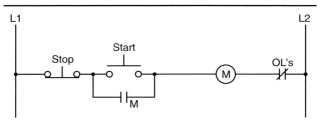

Reprinted from Miller, Charles R. *NFPA's Pocket Electrical References, First Edition.* Jones and Bartlett Publishers, 2007.

COMPLETE STOP-START SYSTEM WITH CONTROL TRANSFORMER

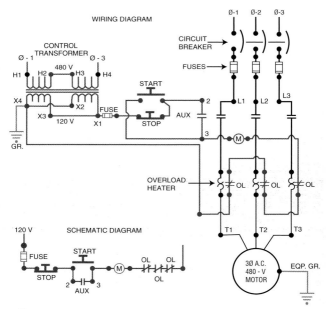

NOTE: Controls and motor are of different voltages.

HAND OFF AUTOMATIC CONTROL

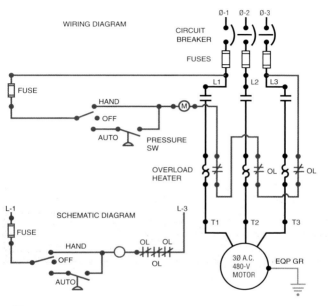

(M) = MOTOR STARTER COIL

NOTE: Controls and motor are of the same voltage.
If Low Voltage controls are used, see page 108 for control transformer connections.

JOGGING WITH CONTROL RELAY

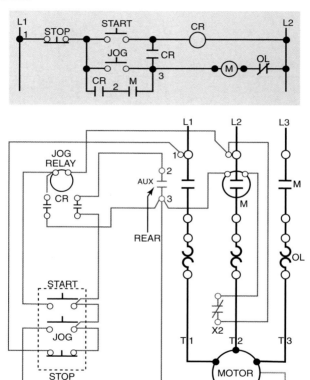

Jogging circuits are used when machines must be operated momentarily for inching (as in set-up or maintenance). The jog circuit allows the starter to be energized only as long as the jog button is depressed.

WIRING DIAGRAMS

Multiple Start and Stop Stations

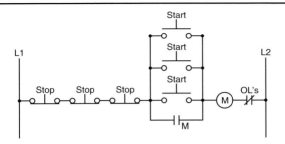

Start Push Button with Job Selector Switch

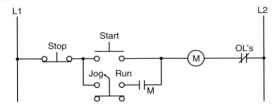

WIRING DIAGRAMS

Reversing Starter with Limit Switches

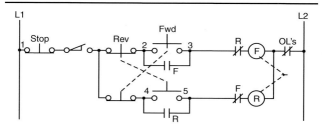

Reversing Starter

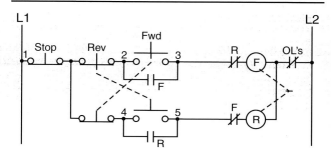

PLUGGING CIRCUIT

Plugging is a method of stopping a motor and load quickly by putting the motor into reverse. A centrifugal switch is necessary to prevent the motor from continuing in reverse.

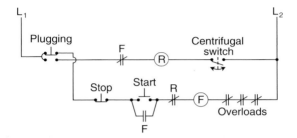

TERMINAL DESIGNATIONS

Generators and Synchronous Motors

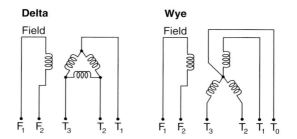

Terminal identifications are the same for both delta and wye, except that the wye arrangement includes a T_0 terminal, which connects to the center-point of the coils.

COUNTER-EMF STARTING

DC Motors

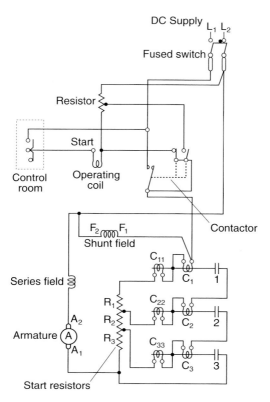

A counter-EMF opposes any change in current flow and is used to reduce high-starting currents.

TWO-SPEED STARTING

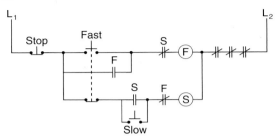

Note this circuit's interlocks:

- Closing the fast switch also opens the Slow leg of the circuit.
- When the S coil is activated, it opens the normally closed S contact in the Fast leg of the circuit.
- When the F coil is activated, it opens the normally closed F contact in the Slow leg of the circuit.

REDUCED-VOLTAGE STARTING

Synchronous Motor

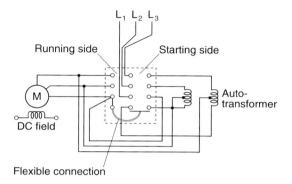

The auto-transformers are in the circuit for starting, and removed afterward.

Ugly's Electrical Reference Series!

The electrical reference series that provides quick, accurate answers whenever and wherever electricians need them.

SERIES TITLES

Ugly's Electrical References
ISBN-13: 978-0-7637-7126-3

Ugly's Referencias Eléctricas
ISBN-13: 978-0-7637-7401-1

Ugly's Electrical Desk Reference, 2008 Edition
ISBN-13: 978-0-7637-7333-5

Ugly's Electrical Safety and NFPA 70E®
ISBN-13: 978-0-7637-6855-3

Ugly's Electric Motors and Controls
ISBN-13: 978-0-7637-7254-3

Ugly's Residential Wiring
ISBN-13: 978-0-7637-7236-9

Special bulk pricing applies to ALL titles in the Ugly's series!

Order Risk-Free Today!
Call: 1-800-832-0034 · Visit: www.jbpub.com

and Controls to the Next Level!

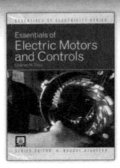

Essentials of Electric Motors and Controls

Charles M. Trout
ISBN-13: 978-0-7637-5113-5
Paperback • 128 Pages • © 2010

This quick, accessible guide is a comprehensive examination of installation and maintenance procedures for motors and controls, as well as a practical introduction to the application and operation of motor control theory. Incorporating numerous illustrations to reinforce key concepts, this book reviews concepts such as magnetism, AC current, frequency, and basic motor operation.

Visit Us Online:

View our full range of electrical products at
www.jbpub.com/electrical

JONES AND BARTLETT PUBLISHERS

Order Risk-Free Today!
Call: 1-800-832-0034 • Visit: www.jbpub.com

The Industry's Only Single-Source Guide to Rotating Machinery!

Stallcup's® Generator, Transformer, Motor and Compressor
2008 EDITION

James G. Stallcup
ISBN-13: 978-0-7637-5255-2
Paperback • 392 Pages • © 2009

This unique training solution provides a complete introduction to all types of rotating machinery; blending basic principles and theory with more advanced operating instructions and maintenance requirements to provide a comprehensive overview of what types of machines are available, how and why each device works, and which Code rules apply in the design and installation.

Visit Us Online:

View our full range of electrical products at
www.jbpub.com/electrical

Order Risk-Free Today!
Call: 1-800-832-0034 • Visit: www.jbpub.com